**Praxiswissen Medizintechnik**

**A guide through the maze of requirements in europe**

# CE-marking
# for medical devices

TÜV Media

M. Becker/S. Menzi/C. Rupprath/S. Schütz

# CE-Marking

Autoren:    M. Becker
            S. Menzl
            C. Rupprath
            S. Scholtz

Preface:    Gerd U. Auffarth

**Bibliografische Information der Deutschen Nationalbibliothek**
Die Deutsche Nationalbibliothek verzeichnet diese Publikation in der Deutschen Nationalbibliografie. Detaillierte bibliografische Daten sind im Internet über http://dnb.d-nb.de abrufbar.

ISBN 978-3-8249-1852-2
© by TÜV Media GmbH, TÜV Rheinland Group, 1. Auflage, Köln 2015
www.tuev-media.de

## Editors and Authors

 Stefan Menzl, Ph.D., serves as Director International Regulatory Affairs for Abbott Medical Optics (AMO). He is in charge of strategic planning, leadership and talent development for AMO´s International Regulatory Departments. Stefan currently manages all aspects of market access and product registration, regulatory compliance and relationship management with international regulatory governing bodies for AMO. In several regions he and his team are involved with obtaining reimbursement for AMO's products. Stefan is a biologist by education and holds a Ph.D. in Microbiology and Biotechnology. He is an active member of the Regulatory Affairs Council of EUCOMED, a member of AdvaMed's European Regulatory Working Group, the BVMed Working Group on Regulatory and Public Affairs (AKRP) and MECOMED's Regulatory Working Group. He has been publishing numerous articles related to Regulatory matters.

 Sibylle Scholtz, Ph.D., biologist and chemist by education, has more than 18 years of experience in the field of medical devices as well as in the field of national and international medical device legislation - at Allergan Medical Optics (AMO) and Abbott Medical Optics. As "Manager Regulatory Affairs EMEA" at AMO she was responsible for several years for the registration process of all medical devices for AMO's "Emerging Markets". Moreover, she has been an active member in several working groups of EUCOMED, MECOMED and EUROMCONTACT, which contributed to her overall understanding and expertise in this field. Sibylle offers more than 22 years of experience in the ophthalmic medical device industry which resulted in high expertise and thorough understanding of developments, processes and future trends.

Carsten Rupprath, Ph.D., is a chemist by education and holds a Ph.D. in Biotechnology. In 2007, Carsten started his career in the medical device industry as "Manager Regulatory Affairs Europe, Middle East, Africa" at Abbott Medical Optics (AMO) with the responsibility to register ophthalmological medical devices within the EMEA region. Since 2012, Carsten is in charge as "Director Regulatory Affairs" of a multinational team with the main tasks of product registration, market access and reimbursement of all AMO products within the EMEA region. He is an active member of the trade associations EUCOMED and EUROMCONTACT.

Myriam Becker has worked in the Medical Device industry for more than 15 years. For about 14 years she was responsible for Medical Device vigilance for Central and South Europe. From 2008 to 2012 she held the position of the deputy Medical Devices Safety Officer (Medizinproduktegesetz/Medical Devices Act § 30). In May 2012 Myriam moved to the Regulatory Affairs department at Abbott Medical Optics where she is responsible for "Regulatory Intelligence" and "review and approval of promotion and training materials" for the region Europe, Middle East and Africa. For more than 10 years she has been an Internal Auditor (Medical Devices); in December 2011 she additionally qualified as QMS Auditor/Lead Auditor (ISO13485:2003).

# Preface

The development of medical technology in the last years of the 20th century has been very dynamic. Synthetic single-use products, artificial joint replacements, pacemaker technologies or minimal-invasive procedures result in a high standard of medical care. But despite of the huge progress in this field, we are presumably only at the beginning of a medical technology revolution.

The 1st version of the German Medizinproduktegesetz (MPG, Medical Devices Act) came into effect on January 1, 1995. It represents the national implementation of the European directives 90/385/EEC for active implantable medical device, 93/42/EEC for medical devices and 98/79/EEC for in vitro diagnostic devices.

With the implementation of this act in the EU internal market, the free movement of medical devices was established. Medical devices that are marketable in one member state of the EU are also marketable in the other EU countries. Manufacturers of medical devices can therefore place their products on the whole EU market.

The CE-mark on the products is the proof of complying with the essential requirements. The Medical Devices Act guarantees the efficacy and safety of the medical device because the product has to undergo comprehensive control procedures during the design and manufacturing process. The CE-mark is a cachet for quality and efficacy. After the market access, the authorities take care of the safety for users and consumers via definition of essential requirements.

This book presents the major aspects of the Medical Devices Act and the European directives. Moreover, the responsibilities and duties of a medical device manufacturer are explained as well as the product classifications and what the manufacturer has to keep in mind when designing and producing medical devices that comply with the legal requirements.

A lot of physicians implant medical devices without being aware of the highly complex registration landscape. Especially in clinical research, the regulations regarding the use of medical devices should be known. The requirements of the ethics committee and of the legal departments of hospitals are getting more complex and therefore especially the scientist should get acquainted with this topic.

Heidelberg, January 2014

Univ.-Prof. Dr. med. Gerd U. Auffarth, FEBO

Medical Devices Consultant

Professor of Ophthalmology

University Hospital Heidelberg, Germany

## Note from the authors

Medical devices are indispensable for the health care sector – in the hospital as well as in the doctor's office. They are used worldwide to save lives, to restore health or prevent the patient's condition from deteriorating. The safety and efficiency of the medical device is therefore of high importance.

The CE-mark that is affixed to medical device is a visible sign that this product meets the essential requirements that are laid down in three major directives that will be discussed later on. This CE-mark is like a "passport" for a medical device that allows the device to "travel" within the European Union. Therefore manufacturers – who want to sell their products in Europe – strive for getting the CE-mark.

This book does not only want to give a theoretical overview of the process of getting the CE-mark and of its importance. It also strives to demonstrate via exercises the practical benefit of this knowledge. You will come acquainted with the major aspects of the European directives and the implementation of these directives into national law. In Germany this law is called Medizinproduktegesetz (MPG, Medical Device Act).

Moreover, you will learn about the responsibilities and obligations of manufacturers of medical devices, about the criteria for product classification, on conformity assessment procedures and many more.

After reading the book and working on the exercises you will be able to decide what method applies to your product. Information on relevant standards is given as well. In the end, you will be acquainted with the major aspects and regulations of the CE-marking process for your medical device.

We hope that you will enjoy the book.

Dr. Stefan Menzl, Dr. Sibylle Scholtz, Dr. Carsten Rupprath, Myriam Becker

PS:

Please keep in mind that all references to laws, directives, standards, guidance documents and websites are as of February 2013. Please always look for the latest revisions of the above mentioned laws, directives etc. As far as standards are concerned, always look for the latest versions of harmonized standards, for there might be more up-to-date standards that are not harmonized yet. For it is still true what Heraclitus of Ephesus once said: "Nothing is as constant as change!"

# Table of Contents

## Chapter 0:  Introduction
*Dr. Stefan Menzl, Dr. Sibylle Scholtz, Dr. Carsten Rupprath, Myriam Becker*

The German gazette "Der Spiegel" once stated that "there was more medical progress in the 20th century than in the whole human history since the Neanderthals ..." and "... there would be no medical progress without the appliance industry ..." (Der Spiegel, 14/1999). In this context, medical devices should be looked at because they are widely used in our health care system today – from prevention, diagnostic to treatment and rehabilitation.

The German Medizinproduktegesetz (MPG, Medical Devices Act)

The 1st edition of the German MPG incorporated the European medical device directive and came into effect on January 1, 1995 (with a transition period until June 14, 1998) and the 4th revision on March 21, 2010. The development of the medical technology landscape has been very dynamic so far. Synthetic single-use products, artificial joint replacements, pacemaker technologies or minimal-invasive procedures result in a high standard of medical care. But despite of the huge progress in this field, we are presumably only at the beginning of a medical technology revolution. The MPG guarantees the safety and efficacy of medical devices. The manufacturers have to comply with the essential requirements in the design and manufacturing process. This compliance results in the CE-mark that is a cachet for quality and efficacy.

What is a Medical Device?

Before a manufacturer gets acquainted with the requirements for a medical device, he has to clarify if his product definitely is a medical device.

The medical device directive 93/42/EEC defines a medical device as follows:

*"... any instrument, apparatus, appliance, software, material or other article, whether used alone or in combination, including the software intended by its manufacturer to be used specifically for diagnostic and/or therapeutic purposes and necessary for its proper application, intended by the manufacturer to be used for human beings for the purpose of:*

- *Diagnosis, prevention, monitoring, treatment or alleviation of disease*
- *Diagnosis, monitoring, treatment, alleviation of or compensation for an injury or handicap*

1

- *investigation, replacement or modification of the anatomy or of a physiological process*
- *control of conception*

*... and which does not achieve its principal intended action in or on the human body by pharmacological, immunological or metabolic means, but which may be assisted in its function by such means ..."*

All these medical devices are regulated in their design, manufacturing and distribution by European directives and regulations as well as by national regulations. The most important European directives are the Medical Device Directive 93/42/EEC from June 14, 1993, the Active Implantable Medical Device Directive 90/385/EEC dated June 20, 1990 and the In Vitro Diagnostic Medical Device Directive 98/79/EEC dated October 27, 1998.

These directives define so-called essential requirements to which a medical device has to comply in order to rightfully bear the CE-mark that enables these products to be distributed in the European market. The national implementations of these directives via the national Medical Devices Act are the basis of placing a product on the market and very often further requirements are added, e. g. language requirements regarding product labelling.

Complying with the essential requirements and the additional requirements of the state law ensures that these products foster health and offer the efficacy promoted by the manufacturer. The CE-mark affixed on the medical device is a synonym for the safety of patients and users.

### The Significance of the CE-mark

By affixing a CE-mark to the label of a medical device, the manufacturer documents that this product complies with all essential requirements. According to the classification of the product and its risk class a so-called Notified Body has to be involved in the assessment of compliance with the essential requirements. In this case, the CE-mark also contains the identification number of the Notified Body.

The procedure that a medical device has to undergo in order to get the CE-mark is called conformity assessment procedure. It contains the following two aspects:

- Safety aspects
  - Analysis, assessment and reduction of risks and adverse events
  - Assessment of the biocompatibility and reduction/elimination of infection risks
  - Electrical safety, electromagnetic safety and mechanical safety
  - Safety in combination with other products
  - Safety and completeness in product labeling and instruction for use
- Efficacy aspects
  - Clinical assessment of medical devices
  - Compliance with all product specifications
  - Proof that product offers the therapeutic or diagnostic benefit that is claimed
  - Guarantee of measurement accuracy

The initial conformity assessment is followed by monitoring the product (and the documentation) as well as the manufacturer and the systems over the whole product life. Every year, the Notified Body conducts audits at the manufacturer and his suppliers. Local surveillance authorities oversee that the manufacturer of medical devices complies with all requirements including post-market surveillance as well as reporting of adverse events.

**Responsibilities**

Who is responsible that a CE-marked medical device definitely complies with all requirements?

- The manufacturer

  According to the medical device directive a "manufacturer means the natural or legal person with responsibility for the design, manufacture, packaging and labeling of a device before it is placed on the market under his own name, regardless of whether these operations are carried out by that person himself or on his behalf by a third party" (MDD 93/42/EEC). By this definition, the manufacturer is not the one who actually produces the device but the one who is responsible for the production ("legal manufacturer"). This "legal manufacturer" is at the end of the day responsible for the compliance with the essential requirements and national requirements. He is the one who affixes the CE-mark on the product and is the "holder" of the CE-certificate that has been issued by the Notified Body. In order to confirm the compliance with the essential requirements, the "legal manufacturer" issues a declaration of conformity.

- Notified Bodies (NB)

Notified Bodies are accredited by the state to conduct the conformity assessment of the manufacturing process as well as whether or not correct and consistent assessment standards were applied. Notified bodies are neutral and independent organizations accredited by the EU member state in which they reside. Organizations outside the EU can be accredited by the commission according to directive 2006/654/EC. That's why there is a Notified Body in Turkey that issues CE-certificates according to MDD 93/42/EEC. The major task of a NB is to carry out conformity assessments of free moving products (if applicable for these products according to the EU directives).

Manufacturers can freely choose a Notified Body. Since January 1, 2010 the e. g. DAkkS (Deutsche Akkreditierungsstelle, German Accrediting Authority) is solely responsible to decide whether or not a Notified Body in Germany – asking to be accredited – is competent and able to conduct the needed conformity assessment and whether or not his independence, impartiality and integrity is beyond doubt. Moreover, the competence of the Notified Body has to be monitored on a regular basis using defined procedures.

The notified bodies assess the quality system and the product of the manufacturer via so-called audits. They conduct type examinations and assess the technical documentation of medical devices.

The Notified Body issues a CE-certificate for a product or a product line when all essential requirements are met. This certificate enables the manufacturer to issue a declaration of conformity and to affix the CE-mark to his product. At this point, the medical device is marketable.

- European representative

  Manufacturers of medical devices that reside outside the EU and want to distribute their products in the European market have to nominate a European representative. This European representative is a natural or legal person and is the interface between the manufacturer and European authorities and users. He is in charge of the major technical documentation and is the contact person for users and state authorities in the markets in which the products are promoted. This representative sometimes also takes care of "vigilance".

- Local competent authorities

  These authorities are ordered by a government of an EU member state to act as defined and are responsible to monitor that all requirements of the three medical device directives (90/385/EEC, 93/42/EEC and 98/79/EEC) are met. Manufacturers and European representatives have to get registered at the authority of the member state in which they have their headquarters. They also have to tell this state authority what medical devices they are planning to place on the market. Beside the surveillance duties, the state authorities have to document and assess all adverse events they receive from manufacturers and users. In some cases they will ask for corrective measures. The state authorities also have to monitor the conformity of non-sterile medical devices of class I (lowest risk class) without measuring function.

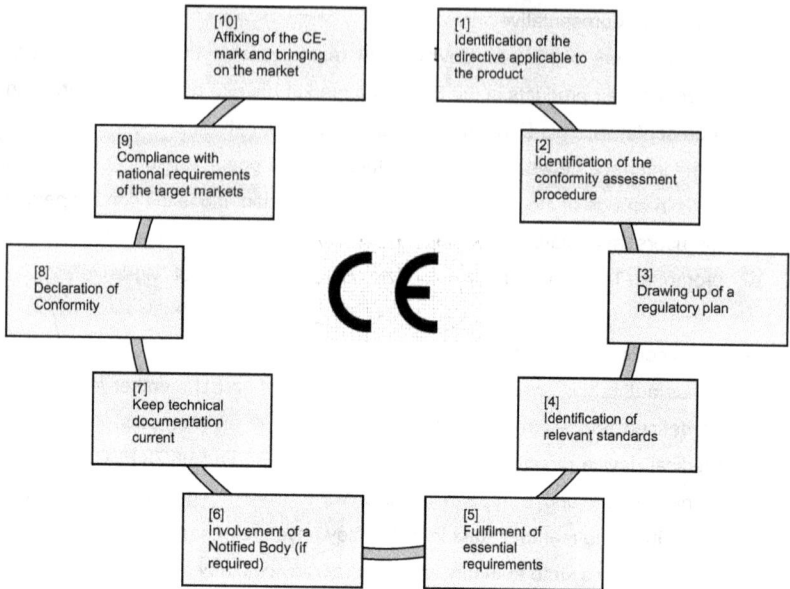

Fig. E/1: The way that leads to the CE mark

In the following chapters the authors will highlight the main topics for you. We hope that you will enjoy this book and that all the information will be of benefit to you.

References:

- Medical Device Directive 93/42/EWG from June 14, 1993
- The Active Implantable Medical Device Directive 90/385/EEC dated June 20, 1990
- The In Vitro Diagnostic Medical Device Directive 98/79/EEC dated October 27, 1998
- Article in the German gazette "Der Spiegel", 14/1999
- Scholtz, S., Freund oder Fallensteller? Das Medizinproduktegesetz (MPG) – seit 1995 in Kraft, Optometrie 2/2008

## Chapter 1: Medical Devices Jurisdiction and European Directives
*Dr. Sibylle Scholtz, Dr. Stefan Menzl*

### 1.1. Learning Objective
In this chapter you will get an overview of the historical development of the medical device legislation and the involved authorities and institutions as well as an understanding of the medical device jurisdiction, of essential requirements for medical devices in Europe and of European directives. You will become acquainted with the major principles of the European legislation, the major legal basis as well as interpretation guidelines.

### 1.2. Historical Background
The European legal system is based on a number of treaties; the most recent one is the Lisbon Treaty from 2009. Article 288 of this treaty defines regulations, directives and decisions as legally binding for all member states whereas recommendations and opinions are not legally binding.

In order to come into effect in the member states, European directives have to be transferred into national law. The member states have a lot of freedom to do so, but the objectives of the directives have to be met. European regulations do not have to be transferred into national law and are therefore directly legally binding for all member states. The European commission usually works on all proposals for European legislation. Finally, the European Parliament and European Council pass the legislation.

The European Commission defines interpretation guidelines, the so-called MEDDEV guidelines. These guidelines are defined by the Medical Device Expert Group (MDEG) and are not legally binding. The European Commission occupies the chair of the MDEG. Moreover, representatives of member state authorities, of the standardization organizations, of Notified Bodies and industry associations are involved as well.
There are also other interpretation guidelines like the NBOG guidelines that are prepared by the Notified Body Operations Group (NBOG).

In case there are several interpretations of a law, the European Commission can be addressed for a consensus. The commission then organizes meetings with representatives of the national authorities and also takes actively part in topic-specific working groups, e. g. for the topic "classification". The European Commission cannot enforce the conformity of national legislation with the European directives. This is the privilege and task of the national authorities.

The Medical Device Directive 93/42/EEC has already been revised five times since 1993. The last revision took place in 2007. The revisions were necessary in order to update the directives on the latest technical developments and to get rid of identified weaknesses in the interpretation of this directive.

The European registration system for medical devices is defined by two principles: the New Approach and the Global Approach. The New Approach allows a faster market entry for products via Europe-wide harmonized directives. The "Global Approach" defines the conformity assessment procedures according to harmonized directives. The "Global Approach" aims at simultaneously recognizing a once performed conformity assessment. Therefore, the "New Approach" and the "Global Approach" complement each other.

Since 1985, the "New Approach" fosters the technical harmonization of defined product groups and the reduction of barriers to trade within the European internal market. These complementary concepts ("New Approach" and "Global Approach") reduce the intervention of the state to a minimum, thus giving the industry a maximum of freedom to meet their obligations. Since 1987 more than 20 EU directives came into effect that are based on the "New Approach" and the "Global Approach".

These harmonized European directives describe the conformity assessment procedure in detail. Products that have successfully passed this procedure can be identified by the CE-mark. The European member states have to transfer the content of EU directives 1:1 into national legislation but can also add further national requirements, e. g. product labelling in the national language.

The legal basis for directives of the "New Approach" is found in Article 95 of the European Council Treaty. The "New Approach" is not limited to medical devices but also covers telecommunication installations, elevators, toys and other products.

As far as medical devices are concerned, Directive 90/385/EEC (The Active Implantable Medical Device Directive) was the first directive implementing the "New Approach". All areas that had already in 1985 Europe-wide legislative character, was not adapted to the "New Approach" (example: pharmaceutical products).

First of all, one has to take care of the classification of a product. The deciding factor for the classification is the way in which a product displays its principal mode of action. The classification might not always be an easy question to decide.

---

**Keep in mind!**

Whereas medicinal products have a pharmacological, immunological or metabolic mode of action, the principal mode of action of a medical device is typically mechanical (MDD 93/42/EEC).

---

There may also be a combination of the above mentioned products but the product has to be classified to one product category only due to its principal mode of action that is defined by the manufacturer.

Harmonized standards play an important role in the European regulation system of medical devices. These standards are recognized by all European member states. By complying with these standards, an assumption of conformity is suggested which means that the product meets all relevant conformity criteria, e. g. of MDD 93/42/EEC. This handling resulted in an easily manageable system for a medical device manufacturer to prove the compliance with the requirements of the directive.

The above mentioned European regulation system for medical devices is based on the following directives:
- Medical Device Directive 93/42/EWG (June 14, 1993)
- The In Vitro Diagnostic Medical Device Directive 98/79/EEC (October 27, 1998)

- The Active Implantable Medical Device Directive 90/385/EEC (June 20, 1990)
- The Personal Protective Equipment Directive 89/686/EEC (December 21, 1989)
- The Animal Tissue Directive 2003/32/EEC (April 23, 2003) with detailed specification to MDD 93/42/EEC
- The Clinical Trial Directive 2001/20/EEC (April 1, 2001)
- The Human Blood and Human Plasma Derivates Directive 2000/70/EEC
- Directive 2003/12/EEC (February 3, 2003) on the reclassification of breast implants
- Directive 2005/50/EEC (August 11, 2005) on the reclassification of hip, knee and shoulder joint replacements
- Directive 2007/47/EC amending Council Directive 90/385/EEC on the approximation of the laws of the Member States relating to active implantable medical devices, Council Directive 93/42/EEC concerning medical devices and Directive 98/8/EC concerning the placing of biocidal products on the market

European member states have to ensure that only products are sold in the European market that do not represent a risk for the safety and health of the population. The member states ensure this via monitoring the conformity of the products with the essential requirements of a defined directive as well as via post-market surveillance.
The essential requirements are defined in the annex of a directive. By complying with these essential requirements, medical devices are not only safe for patients, users and others but also as efficient as claimed by the manufacturer. Moreover, the manufacturer has also to keep in mind the latest technical developments and the day-to-day practice.

A directive describes in detail the process of assessing the conformity that a manufacturer has to undergo before affixing the CE-mark to the product and placing it on the market. In the majority of these conformity assessment processes (depending on the risk class of the medical device), a Notified Body is involved.
Notified Bodies are accredited by the state to conduct the conformity assessment of the manufacturing process as well as whether or not correct and consistent assessment standards were applied. Manufacturer can freely choose a Notified Body that is accredited for a defined procedure or specified product category. The service

portfolio can be different from one Notified Body to another but all Notified Bodies are independent and private-sector organizations. At the moment there are 75 Notified Bodies in Europe.

---

**Exercise:**

Visit the homepage of the European Commission and print out the list of Notified Bodies regarding Directive 93/42/EEC.

---

Notified Bodies have to meet special requirements in order to get accredited. These requirements are:

- Quality system
- Compliance with the accreditation regulations
- Neutrality, impartiality and independence
- Competent staff

The compliance with these requirements by the Notified Body is monitored by the Competent Authority of the EU member state.

The conformity assessment procedure for a medical device depends mainly on its risk class classification. There are four risk classes (I, IIa, IIb, III).The higher the risk class, the higher the potential risk of the medical device.

A manufacturer of a medical device of risk class I can conduct the conformity assessment procedure without involving a Notified Body. But only in case the medical device does not have a measurement function and is not sterile.

In the conformity assessment procedures of medical devices of risk class IIa, IIb and III, a Notified Body has to be involved. The Notified Body carries out audits and examines the technical documentation. After successfully conducting the conformity assessment procedures, the Notified Body issues a so-called CE-certificate that is followed by a declaration of conformity by the manufacturer. Then the CE-mark is affixed to the medical device. The CE-mark is like a "passport" that enables the product to "travel" within the European Union – with only a few limitations that will be addressed later on.

## 1.3. Authorities, Institutions and other Relevant Players

### 1.3.1. Learning Objective

In this part information on the most important institutions for the European registration system of medical devices are given as well as on the major players of the CE-marking process and what role and responsibilities these major players have.

### 1.3.2. The European Commission, Parliament & Council

As already mentioned the European Commission proposes and works on the content of laws. Decisions then have to be implemented in these laws. The proposal is introduced to the European Parliament and European Council who finally passes the bill (or asks for modifications). The European Commission has 25.000 staff members and is organized in so-called Directorates General. The Directorate that is responsible for medical devices (as well as for medicinal products) is the Directorate General for Health and Consumer Protection (DG SANCO).

---

**Exercise:**

Go to the homepage of the European Commission and look for the structure of the Directorates.

---

Although the European Commission can institute proceedings because of violation of a law following a complaint, the control of the implementation of legislation is the obligation of the EU member states. Therefore, the legislation of the individual EU country is relevant for the manufacturer of medical devices.

### 1.3.3. National Legal Institutions and Authorities

National legislation (laws, regulations, directives) of the country in which the manufacturer is residing, is of major importance because national requirements define the rights and duties of legal manufacturers, EU representatives as well as manufacturing sites. When a manufacturer wants to place medical devices on a market, the national requirements of that target country are key.

The relevant laws of the EU member states do not differ much because they are derived from European directives, regulations and decisions. Nevertheless, the member states are free to add additional requirements in order to define areas that have not been described in detail (e. g. the obligation to inform about marketed medical devices, requirements regarding native language product labelling).

### 1.3.4. Notified Bodies

Notified Bodies are accredited by the EU member states. The accreditation is then reported to the EU Commission. It is the Notified Body´s task to verify the conformity assessment of the medical device on behalf of the manufacturer using uniform assessment criteria and issuing a CE-certificate.

Manufacturer can freely choose a Notified Body that is accredited for a defined procedure or specified product category. The service portfolio can be different from one Notified Body to another but all Notified Bodies are independent and private-sector organizations. At the moment there are 75 Notified Bodies in Europe (list of accredited Notified Bodies can be found on the homepage of the EU Commission: http://ec.europa.eu/enterprise/newapproach/ nando/).

---

**Exercise**

Look for two Notified Bodies in Germany that are accredited for orthopaedic implants.

---

Notified Bodies are organized in the European Forum of Notified Bodies Medical Devices (NB-MED). They publish recommendations on the implementation of European directives on a regular basis. These recommendations are the result of the exchange of the Notified Bodies with one another as well as with representatives of industry associations and the EU Commission. These recommendations are not binding. They help to interpret European directives for manufacturers, Notified Bodies and others.

---

**Exercise:**

Look up the latest list of NB-MED guidelines.

---

TEAM-NB (www.team-nb.org) is the abbreviation for the European Association for Medical Devices of Notified Bodies. This is a non-profit organization of Notified Bodies according to the following three directives 90/385/EEC, 93/42/EEC and 98/78/EC.

The objectives of this organization are as follows:

- Foster the communication between EU Commission, industry, responsible authorities and users
- Communicate consolidated opinion of all concerned Notified Bodies
- Support a high technical and ethical standard regarding the work and service of Notified Bodies
- Protection of legal and commercial interests of Notified Bodies

TEAM-NB was founded in 2001 and consists of 32 members. All NB-MED recommendations are published on the following website http://team-nb.org.

## 1.3.5. Standardization Organizations

At the moment there are three relevant standardization organizations in Europe:

- The European Committee for Standardization (CEN)
- The European Committee for Electrotechnical Standardization (CENELEC)
- The European Telecommunications Standards Institute (ETSI)

CEN develops standards for non-active medical devices (www.cen.eu). CENELEC focuses on active (electrically powered) medical devices (www.cenelec.eu) and ETSI takes care of standards for information technology and telecommunications technology (www.etsi.org).

All these organizations develop standards on a European level. There are also other standardization organizations, e. g. consortia of manufacturers and users that define standards for specific technology areas. Standards are important success factors.

## 1.3.6. Importers and Distributors

In every country there are different requirements for the authorization to import and distribute medical devices. Moreover, importers and distributors have defined duties regarding the correct storage and transport of medical devices (according to the labelling by the manufacturer). Distributors have to be able to trace back medical devices to the next customer. This is of great importance in case of recalls or the application of safety-relevant measures.

## 1.3.7. User of Medical Devices

Rights and duties of users of medical devices are not defined in the directives but on a national level. The duties refer to a great extent to the reporting of adverse events and as far as health care organizations are concerned the traceability of the product to the individual patient.

---

**Exercise:**

Take the German Medizinproduktegesetz (Medical Devices Act) and look for the definition of rights and duties of users of medical devices.

---

## 1.3.8. The Manufacturer

According to the medical device directive (MDD 93/42/EEC) a "manufacturer means the natural or legal person with responsibility for the design, manufacture, packaging and labeling of a device before it is placed on the market under his own name, regardless of whether these operations are carried out by that person himself or on his behalf by a third party". The manufacturer defines the intended use of the medical device. Moreover, it is his responsibility that his products comply with the essential requirements. Before affixing the CE-mark, he signs a declaration of conformity. The manufacturer has to set up the technical documentation and keep it up-to-date. He has to establish a vigilance system and to report adverse events, to assess these events and to implement safety-relevant measures if needed.

A manufacturer of medical devices of class I has to

- Decide whether his product is a medical device or an accessory of a medical device according to the intended use
- Go through the classification criteria in Annex IX of the directive in order to make sure that his product definitely is a class I product
- Make sure that his product complies with the essential requirements (Annex I of the directive)
- Establish a technical documentation of his product (also for the inspection by the Competent Authority)
- Sign the declaration of conformity before affixing the CE-mark
- Establish and maintain a monitoring system

- Involve a Notified Body in case his product is sterile or has a measurement function
- Inform the Germany Competent Authority about first making available of a product in specific countries

As far as medical devices of class IIa, IIb and III are concerned the manufacturer closely works together with his Notified Body of choice.

### 1.3.9. The European Representative

The manufacturer is responsible for the first making available of a medical device. If the manufacturer does not reside in Europe, he has to appoint a European representative or an authorized person. If no authorized person is appointed or if medical devices are not imported to the EU on behalf of the authorized person, the importer is responsible.

The European representative or authorized person is a natural or legal person residing in Europe that was appointed by the manufacturer to act on his behalf and to be available for the Competent Authorities. In that case, the name and address of the manufacturer as well as the name and address of the representative have to be listed on the product label or instruction for use of the medical device.

The responsibilities of the European representative have been interpreted by the individual member states in a different way, e. g. that the European representative is only the spokes-man of the manufacturer towards the authorities of the EU member states or even that the European representative has the same responsibilities as the manufacturer.

In January 2012, the European Commission published a new guideline MEDDEV 2.5/10, the Guideline for Authorized Representatives) regarding the tasks of a European representative: this guideline defines the function and responsibilities as well as the expectations of the member states regarding post-market surveillance. The main focus is on a draft agreement between a manufacturer and the European representative that defines the function and responsibilities for the representative and the obligations of the manufacturer.

It has to be guaranteed that the European representative is competent and has access to the technical documentation. He also has to be available for the Competent Authorities and to answer questions if needed. The Directive 93/42/EEC (amended by Directive 2007/47/EEC) defines that a manufacturer is only allowed to appoint one European representative for a medical device. The European representtative has to inform the responsible national authority about his designation. Some European representatives are members of the European Association of Authorized Representatives (www.eaarmed.org).

## 1.4. Summary

By affixing the CE-mark to his products, a manufacturer documents the conformity with the essential requirements. According to the risk classification of the product, a Notified Body has or has not to be involved. The identification number of the Notified Body hast to be added to the CE-mark. The compliance with all requirements is proven via a conformity assessment procedure that consists of:

- Safety aspects
    - o Analysis, assessment and reduction of risks and adverse events
    - o Assessment of the biocompatibility and reduction/elimination of infection risks
    - o Electrical safety, electromagnetic safety and mechanical safety
    - o Safety in combination with other products
    - o Safety and completeness in product labelling and instructions for use
- Efficiency aspects
    - o Clinical assessment of medical devices
    - o Compliance with all product specifications
    - o Proof that product offers the therapeutic or diagnostic benefit that is claimed
    - o Guarantee of measurement accuracy

This has to be guaranteed over the whole product life!

General information on the German Medizinprodukterecht (MPG, Medical Devices Act) and a version of the revised MPG are published on the following websites: www.bvmed.de, http://bundesrecht.juris.de/mpg/index.html or www.dimdi.de

## 1.5. Test Your Knowledge

**Q1:** A manufacturer of medical devices with headquarter and production sites in the US wants to distribute his products in Europe. What central function does this manufacturer have to appoint in order to be able to do so?

**A1:** He has to appoint a European representative.

---

**Q2:** How can a manufacturer find out which authority is responsible for him or for a defined topic? Is this answer given in the MDD Directive 93/42/EEC?

**A2:** The responsibility is defined in national laws (in Germany: state or federal authorities). No, the MDD Directive 93/42/EEC does not answer this question.

---

**Q3:** Which of the following statements is true?

1. A national responsible authority controls a Notified Body.

2. A Notified Body has to be independent of a manufacturer.

3. The Notified Body can be a private-sector organization.

**A3:** All three answers are true.

---

**Q4:** What are the tasks of a Notified Body?

1. Conduction of audits regarding quality management on a regular basis

2. Approval of essential changes of the certified quality management system

3. Approval of the quality management system

4. Approval of all changes of the certified quality management system

**A4:** Answers 1 and 2 are correct.

---

**Q5:** Example: When defining the intended use of a product, manufacturer and Notified Body disagree. Who in the end has the right to decide?

**A5:** The manufacturer.

## 1.6. References

- Lisbon Treaty: http://europa.eu/lisbon_treaty/index_de.htm
- MEDDEV Guidelines: http://ec.europa.eu/health/medical-devices/documents/guidelines/index_en.htm
- www.meddev.info
- NB-MED: www.team-nb.org/
- 93/42/EWG: http://eur-lex.europa.eu/LexUriServ/LexUriServ.do?uri=CONSLEG:1993L0042:20071011:de:PDF
- 90/385/EWG, http://eur-lex.europa.eu/LexUriServ/LexUriServ.do?uri=CONSLEG:1990L0385:20071011:de:PDF
- 98/79/EC, http://eur-lex.europa.eu/LexUriServ/LexUriServ.do?uri=CELEX:31998L0079:en:NOT

# Chapter 2: Conformity Assessment

*Dr. Stefan Menzl*

## 2.1. Learning Objective

In this chapter the various conformity assessment procedures for medical devices are discussed one of which the manufacturer has to apply. You will get acquainted with the special features of the procedures and the advantages of a defined procedure for a medical device.

## 2.2. Conformity Assessment – an Overview

The conformity assessment of a medical device depends on the EU directive by using one out of several versions of conformity assessment procedures due to the risk classification of the medical device. The procedure assesses whether or not a medical device complies with essential requirements.

The essential requirements consist of three topics:

- Safety
- Technical performance
- Medical performance

Regarding the EU directives, there are various procedures that can be used to examine and assess the conformity of the products. The procedures depend on the classification of the medical device. There are procedures for every classification and the manufacturer can choose the one that makes the most sense to him.

## Conformity Assessment Procedure 93/42/EEC

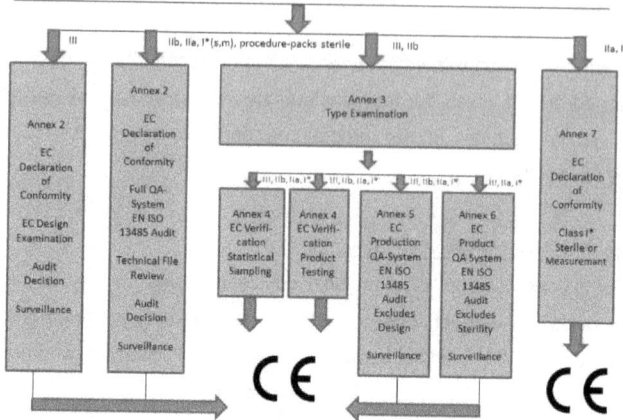

Fig. 2/1: Overview Options Conformity-Assessment-Procedures

When using the procedure that is defined in Annex VII of the directive 93/42/EEC, only the manufacturer assesses the conformity of his products. A notified body is not involved.

When using Annex VI, the manufacturer has to implement a quality management system. This system defines the final inspection of the manufactured products. The design phase and the production phase are not covered. The quality management system is checked and certified by a notified body that conducts regular audits.

When using Annex V, the manufacturer establishes a quality management system that describes the production and final inspection of the product, but not the design phase. The quality management system is checked and certified by a notified body that conducts regular audits.

In the procedure defined in Annex IV, the manufacturer does not establish a quality management system. The products are examined by the notified body after production. The notified body can examine all products or verifying chosen lots or serial numbers.

When using the procedure defined in Annex III of the directive 93/42/EEC, a representative product sample is taken from the production and will then be examined by the notified body whether it complies with the essential requirements. This procedure is complemented by the examination of the design dossier. This procedure is called EU type examination. The certification only refers to products of the same type or products that were produced in identical processes.

A conformity assessment due to Annex II results in the manufacturer establishing a quality management system. This system describes all phases: design, production, final inspection and product release. A notified body examines and certifies the quality management system and conducts audits on a regular basis.
For class III devices, the manufacturer sets up a design dossier for a medical device and forwards this design dossier to a notified body. The notified body then checks this dossier.

## Conformity Assessment Procedure 93/42/EWG

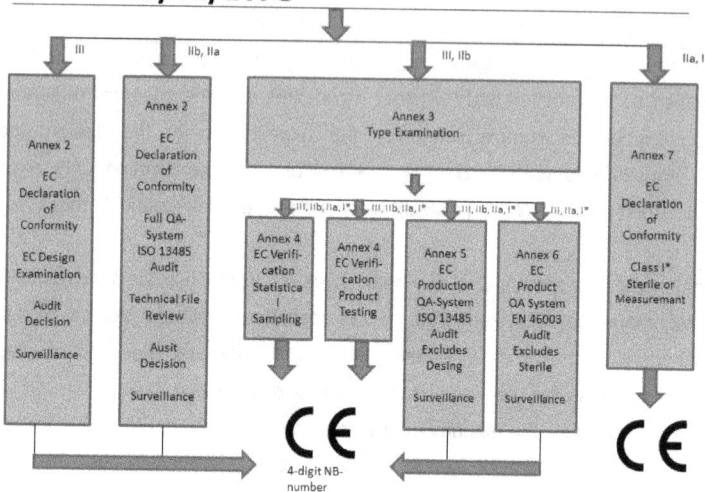

Fig. 2/2: Conformity Assessment Procedures 93/42/EWG

When using Annex I, the manufacturer of the medical device assures that his products meets all essential requirements. It can be presumed that a product complies with the essential requirements when harmonized standards were used.

It is the sole responsibility of a manufacturer that his products comply with the essential requirements of the defined EU directive. The notified body assesses the conformity of the quality management system with the requirements of the EU directives (Annex II, V, VI).

Exceptions are e. g. medical devices incorporating – as an integral part – a substance which (if used separately) can be considered to be a medicinal product, medical devices with tissue of animal origin or medical devices consisting of human blood or human plasma derivates. These exceptions are covered by a separate chapter of this book.

## 2.3. Conformity Assessment – Essential Requirements

In order to assess conformity of medical devices, the essential requirements for each directive (93/42/EEC MDD, 90/385/EEC AIMDD) are specified in Annex I. In order to comply with these requirements, so-called harmonized standards can be applied.

If the manufacturer does not apply harmonized standards although these exist, the manufacturer has to give detailed reasons to justify his decision. Cave: The chosen alternative has to have at least the same safety level!

A key element of the conformity assessment of medical devices is related to the technical documentation. This documentation consist amongst other elements of

- Classification of the product
- Proof of the compliance with the essential requirement according to Annex I
- Risk analysis (EN ISO 14971)
- Safety testing
- Clinical assessment

General conditions to fulfil the essential requirements regarding
Directive 93/42/EEC (1-6) are:

- No risks which compromise the clinical condition or safety including ergonomics taking intended users into consideration
- Conformity concerning safety considering state-of-the-art technical knowledge, fulfilment of a specified performance, characteristics and performance have to be maintained over a specified product life, as well during storage and transportation
- Benefit has to outweigh risks, proof of clinical assessment according to Annex X

The essential requirements to design and construction according to Directive 93/42/EEC (7-12) consist of:

- Chemical, physical and biological properties
- Infection and microbial contamination
- Construction and environmental properties
- Devices with a measuring function
- Protection against radiation

- Requirements for medical devices connected or equipped with an energy source

Further essential requirements are information that has to be supplied by the manufacturer (Directive 93/42/EEC (13):
- Product description
- Product packaging/labelling
- Instruction for use
- Requirement to set up a technical documentation that should be always up-to-date and provided for inspection by the Competent Authorities if needed
- Risk management file for each product
- Clinical assessment of each product
- Compliance with the essential requirements has to be documented
- A declaration of conformity must be issued for every product that is placed on the market

## 2.4. Establishing a Technical Documentation

The following parts have to be included in a technical documentation:
- Introduction
- Essential requirements checklist (ERCL)
- Risk analysis
- Design and product specifications
- Chemical, physical and biological tests
  - Bench-testing, pre-clinical studies
  - Biocompatibility tests
  - Biostability tests
  - Microbiological safety, tissue of animal origin
- Clinical data
- Packaging qualification and shelf life
- Labelling and instruction for use, patient information, promotional material
- Production
- Sterilisation

- Summary
- Declaration of conformity

As far as the conformity assessment of medical devices is concerned, the risk analysis is one of the most important parts of the technical documentation. According to ISO 14971, an adequate risk management consists of a systematic application of management policies, procedures and practices to the tasks of analysing, evaluating and controlling risk (ISO 14971, § 2.18). All these activities are documented in the risk management file.

The structure of an adequate risk analysis could be as follows:
- Introduction (short description of the product, product history, market experience, incidents, number of cases in relation to units placed on the market)
- Risk analysis over the product life cycle, monitoring of the product over the life cycle
- Questions regarding the definition of the product features (applicable, not applicable)
- Classification of risks (occurrence probability, severity, risk acceptance)

## Riskanalysis

| Place where Harm occurs | Harm | Severity of Harm | Probability of Occurence of Harm | Risk | Risk-Reduction | Remaining Risk | Risk-Acceptance |
|---|---|---|---|---|---|---|---|
| Component of Product incl. Packaging & Labelling | | | | | | | |

| Product -Use | Harm | Severity of Harm | Probability of Occurence of Harm | Risk | Risk-Reduction | Remaining Risk | Risk-Acceptance |
|---|---|---|---|---|---|---|---|
| Product | | | | | | | |
| Defects in the Product | | | | | | | |
| Unintended use of the Product | | | | | | | |

Fig. 2/3: Example of Structure for Risk-Analysis

Ideally the result of the risk analysis should classify the use risk of the product as "low" that means manageable (compare MDD 93/42/EEC, Annex I, Essential requirements, general requirements; reference: EN ISO 14971).

As far as the form of the risk analysis is concerned, the document must be dated and signature by the responsible person.

A clear structure and a better overview of the risk analysis can be achieved by using checklists that contain the essential requirements. The proof of the compliance with the essential requirements can be seen at a glance and also the reference to the sources of the individual proofs in the technical documentation as well as to the applied standards.

**Keep in mind!**
Complying with standards is voluntary. But meeting these standards prevents developing own test specification and having discussion about the state-of-the-art.

Complying with harmonized European standards results in the presumption of conformity with the essential requirements (Directive 93/42/EEC, Article 5.1*).

\* Member States shall presume compliance with the essential requirements referred to in Article 3 in respect of devices which are in conformity with the relevant national standards adopted pursuant to the harmonized standards the references of which have been published in the Official Journal of the European Communities; Member States shall publish the references of such national standards.

When applying standards the hierarchical structure should be considered:

- Harmonized standards (published in the official journal of the EU Commission)
- European standards
- International standards (EC, ISO, etc.)
- EU National standards (DIN, BSI, etc.)
- Third party standards (ASTM, AAMI)
- Manufacturer specifications

Below, please find an example of a part of an essential requirements checklist:

| | ESSENTIAL REQUIREMENTS | STANDARDS USED | EVIDENCE OF COMPLIANCE |
|---|---|---|---|
| 12.8.1 | Devices for supplying the patient with energy or substances must be designed and constructed in such a way that the flow-rate can be set and maintained accurately enough to guarantee the safety of the patient and of the user. | EN 60601-1<br><br>IEC 60601-2-24<br><br>ANSI/AAMI ID 26: Infusion Devices | 142-36 – 601-1 Test Report by UL<br>142-37 – 602-2-24 Test Report by BSI<br><br>142-72 – ID 26 Test Report |
| 13.6 | Where appropriate, the instructions for use must contain the following particulars: *(list omitted for brevity here but details should be included)* | EN 60601-1 (clause 6)<br><br>IEC 60601-2-24<br><br>SOP-87: Instructions for Use | 142-36 – 601-1 Test Rpt by UL<br>142-37 – 602-2-24 Test Report by BSI<br>142-89 – IFU Verification Report |

Fig. 2/4: Excerpt of an Essential Requirements Checklist

## 2.5. Summary

The term "conformity assessment" defines a procedure to prove the compliance of the product with the legal requirements (essential requirements). Please also compare with chapter 4.

## 2.6. Test Your Knowledge

| | |
|---|---|
| **Q1:** | Why is the use of harmonized standards important? |
| **A1:** | Because using harmonized standards results in the presumption of conformity. |

| | |
|---|---|
| **Q2:** | What should the manufacturer use as a basis of the conformity assessment? |
| **A2:** | a) the intended use of the medical device and |
| | b) standards (e. g. harmonized standards, product-specific standards) |

| | |
|---|---|
| **Q3:** | A harmonized standard for a medical device exists but the manufacturer does not want to apply this standard. Is he able to prove the compliance with the essential requirements nevertheless? |
| **A3:** | Yes. Applying harmonized standards is not mandatory. But the manufacturer has to prove that his product meets the essential requirements. |

| | |
|---|---|
| **Q4:** | A safety-relevant, product-specific standard for a medical device has been revised. The manufacturer has designed his product according to the now "old version" of this standard. Does the manufacturer have to re-design his product in order to prove the compliance with the essential requirements? |
| **A4:** | The manufacturer has to make sure that his product meets the safety-relevant standards. He may have to re-design his product. The first step is to conduct a "gap analysis" of the old versus the modified standard. He then assesses the deviating elements regarding the risk followed by setting up a plan how to comply with the new requirements. This plan should be discussed with the Notified Body resulting in the decision whether or not a re-design is necessary. |

**Q5:** What statement concerning harmonized standards is correct?

1. Using harmonized standards usually results in authorities presuming compliance with legal requirements

2. Harmonized standards are approved by the European Standardization Organisations CEN, CENELEC and/or ETSI

3. The application of harmonized standards is voluntary.

**A5:** All 3 answers are correct.

---

**Q6:** What institution has to be involved in a conformity assessment procedure?

1. The manufacturer

2. The BfArM (Federal Institute for Drugs and Medical Devices)

3. The notified body

4. The responsible Regierungspräsidium (Administrative District Council)

**A6:** Answer 1 is correct

---

**Q7:** What statement about the intended use of a medical device is true?

1. The manufacturer and the notified body jointly decide on the intended use.

2. The intended use is defined by a possible use.

3. The Regierungspräsidium (Administrative District Council) has to approve the intended use.

4. The manufacturer defines the intended use. The decision, however, must be taken in an objective way.

**A7:** Answer 4 is correct.

---

**Q8:** Where can you find the list of essential contents of the technical documentation?

**A8:** In the Annex II – VII of the MDD 93/42/EEC

## 2.7. References

- www.bfarm.de/DE/Medizinprodukte/inverk/inverk-node.html
- www.medcert.de/konformitaetsbewertung
- Conformity assessment (ZLG)
  www.zlg.de/index.php?eID=tx_nawsecuredl&u=0&file=fileadmin/downloads/ab/309_0406_A07.pdf&hash=ce4f1fde2f17068b24f4c24c4eca3be0565baa07

- MDD 93/42/EEC
  http://eur-
  lex.europa.eu/LexUriServ/LexUriServ.do?uri=CELEX:31993L0042:DE:NOT
- MEDDEV
  http://ec.europa.eu/health/medical-
  devices/documents/guidelines/index_en.htm
- Notified Body Operation Group
  www.nbog.eu/
- Directive 2003/32/EC with specifications regarding 93/42/EEC defined
  requirements for medical devices with tissue of animal origin http://eur-
  lex.europa.eu/LexUriServ/LexUriServ.do?uri=OJ:L:2003:105:0018:0023:DE:P
  DF
- Guidance Notes for Manufacturers of Class I Medical Devices endorsed by the
  MDEG on December 2009
  http://ec.europa.eu/health/medical-devices/files/guide-stds-directives/notes-
  for-manufacturers-class1-09_en.pdf
- Guidance Notes for Manufacturers of Custom-Made Medical Devices
  endorsed by the MDEG on June 2010
  http://ec.europa.eu/health/medical-devices/files/guide-stds-directives/notes-
  for-manufactures-custom-made-md_en.pdf

# Chapter 3: Product Classification and Rules of Classification

*Dr. Stefan Menzl, Dr. Sibylle Scholtz*

## 3.1. Learning Objective

In this chapter you will learn about the risk classes of medical devices and you will finally be able to classify your products by yourself. The classification according to risk classes is a central part of the assessment of a medical device and influences the conformity assessment as well as the quality of the technical documentation.

## 3.2. Introduction

The classification of medical devices is described in the following directives:

- 90/385/EEC AIMD
- 93/42/EEC MDD

---

**Keep in mind!**

Before starting with the classification you should make sure that this directive definitively applies.

---

The classification depends on the potential risk of the medical device. One has to assess the following topics:

- vulnerability of the human body
- risks regarding the design of a product
- risks during the manufacturing process.

## 3.3. Medical Device – Definition

Medical devices are – e. g. according to §3 of the German MPG (Medical Devices Act) – "all instruments, apparatus, appliances, software, substances or preparations made from substances or other articles, used alone or in combination, including the software intended by the manufacturer to be used specifically for diagnostic or therapeutic purposes and necessary for the medical device's proper application, intended by the manufacturer to be used for human beings by virtue of their functions, for the purpose of ...

- Diagnosis, prevention, monitoring, treatment or alleviation of disease
- Diagnosis, monitoring, treatment, alleviation or compensation of injuries or handicaps
- Investigation, replacement or modification of the anatomy or of the physical process
- Control of conception

... and which do not achieve their principal intended action in or on the human body by pharmacological, immunological or metabolic means but which might be assisted in their function by such means. Products that achieve their principal intended action in or on the human body by pharmacological, immunological or metabolic are medicinal products".

### 3.4. Risk Classes

In the MDD 93/42/EEC the following risk classes are described:

- Class I
- Class IIa
- Class IIb
- Class III

Class I can be modified by the product specification "sterile": class Is or by the product specification "with measuring function": class Im.

In the MDD 93/42/EEC there are also described special procedures that are conducted independently of the classification:

- Custom-made products
- Medical devices for clinical investigation

The 18 rules of classifications are to be found in Annex IX of the MDD 93/42/EEC and describe and assess ...

- Duration
  - Transient (< 1h), short term (< 30 days), long term (> 30 days)

- Invasive devices
  - Invasive device, body orifice, surgically invasive device, implantable device
- Place of application
  - Central circulatory system, central nervous system, direct contact with the heart, other
- Energy source (active, non-active)

In the following part, the rules are described in detail:

## NON-INVASIVE DEVICES

### Rule 1

All non-invasive devices are in Class I, unless one of the rules set out hereinafter applies.

### Rule 2

All non-invasive devices intended for channelling or storing blood, body liquids, or tissues, liquids or gases for the purpose of eventual infusion, administration or introduction into the body are in Class IIa:

- If they may be connected to an active medical device in Class IIa or a higher class,
- If they are intended for use for storing or channelling blood or other body liquids or for storing organs, parts of organs or body tissues, in all other cases they are in Class I.

### Rule 3

All non-invasive devices intended for modifying the biological or chemical composition of blood, other body liquids or other liquids intended for infusion into the body are in Class IIb, unless the treatment consists of filtration, centrifugation or exchanges of gas, heat, in which case they are in Class IIa.

### Rule 4

All non-invasive devices which come into contact with injured skin:

- Are in Class I if they are intended to be used as a mechanical barrier, for compression or for absorption of exudates,

- Are in Class IIb if they are intended to be used principally with wounds which have breached the dermis and can only heal by secondary intent,
- Are in Class IIa in all other cases, including devices principally intended to manage the micro-environment of a wound.

## INVASIVE DEVICES

### Rule 5

All invasive devices with respect to body orifices, other than surgically invasive devices and which are not intended for connection to an active medical device or which are intended for connection to an active medical device in Class I:

- Are in Class I if they are intended for transient use,
- Are in Class IIa if they are intended for short-term use, except if they are used in the oral cavity as far as the pharynx, in an ear canal up to the ear drum or in a nasal cavity, in which case they are in Class I,
- Are in Class IIb if they are intended for long-term use, except if they are used in the oral cavity as far as the pharynx, in an ear canal up to the ear drum or in a nasal cavity and are not liable to be absorbed by the mucous membrane, in which case they are in Class IIa. All invasive devices with respect to body orifices, other than surgically invasive devices, intended for connection to an active medical device in Class IIa or a higher class, are in Class IIa.

### Rule 6

All surgically invasive devices intended for transient use are in Class IIa unless they are:

- Intended specifically to control, diagnose, monitor or correct a defect of the heart or of the central circulatory system through direct contact with these parts of the body, in which case they are in Class III,
- Reusable surgical instruments, in which case they are in Class I,
- Intended specifically for use in direct contact with the central nervous system, in which case they are in Class III,
- Intended to supply energy in the form of ionising radiation in which case they are in Class IIb,
- Intended to have a biological effect or to be wholly or mainly absorbed in which case they are in Class IIb,

- Intended to administer medicines by means of a delivery system, if this is done in a manner that is potentially hazardous taking account of the mode of application, in which case they are in Class IIb.

Rule 7

All surgically invasive devices intended for short-term use are in Class IIa unless they are intended:

- Either specifically to control, diagnose, monitor or correct a defect of the heart or of the central circulatory system through direct contact with these parts of the body, in which case they are in Class III,
- Or specifically for use in direct contact with the central nervous system, in which case they are in Class III,
- Or to supply energy in the form of ionizing radiation in which case they are in Class IIb,
- Or to have a biological effect or to be wholly or mainly absorbed in which case they are in Class III,
- Or to undergo chemical change in the body, except if the devices are placed in the teeth, or to administer medicines, in which case they are in Class IIb.

Rule 8

All implantable devices and long-term surgically invasive devices are in Class IIb unless they are intended:

- To be placed in the teeth, in which case they are in Class IIa,
- To be used in direct contact with the heart, the central circulatory system or the central nervous system, in which case they are in Class III,
- To have a biological effect or to be wholly or mainly absorbed, in which case they are in Class III,
- Or to undergo chemical change in the body, except if the devices are placed in the teeth, or to administer medicines, in which case they are in Class III.

## ADDITIONAL RULES APPLICABLE TO ACTIVE DEVICES

Rule 9

All active therapeutic devices intended to administer or exchange energy are in Class IIa unless their characteristics are such that they may administer or exchange energy to or from the human body in a potentially hazardous way, taking account of the nature, the density and site of application of the energy, in which case they are in Class IIb.

All active devices intended to control or monitor the performance of active therapeutic devices in Class IIb, or intended directly to influence the performance of such devices are in Class IIb.

### Rule 10

Active devices intended for diagnosis are in Class IIa:

- If they are intended to supply energy which will be absorbed by the human body, except for devices used to illuminate the patient's body, in the visible spectrum,
- If they are intended to image *in vivo* distribution of radiopharmaceuticals,
- If they are intended to allow direct diagnosis or monitoring of vital physiological processes, unless they are specifically intended for monitoring of vital physiological parameters, where the nature of variations is such that it could result in immediate danger to the patient, for instance variations in cardiac performance, respiration, activity of CNS in which case they are in Class IIb. Active devices intended to emit ionizing radiation and intended for diagnostic and therapeutic interventional radiology including devices which control or monitor such devices, or which directly influence their performance, are in Class IIb.

### Rule 11

All active devices intended to administer and/or remove medicines, body liquids or other substances to or from the body are in Class IIa, unless this is done in a manner:

- That is potentially hazardous, taking account of the nature of the substances involved, of the part of the body concerned and of the mode of application in which case they are in Class IIb.

### Rule 12

All other active devices are in Class I.

## SPECIAL RULES

### Rule 13

All devices incorporating, as an integral part, a substance which, if used separately, can be considered to be a medicinal product, as defined in Article 1 of Directive 2001/83/EC, and which is liable to act on the human body with action ancillary to that of the devices, are in Class III.

All devices incorporating, as an integral part, a human blood derivative are in Class III.

Rule 14

All devices used for contraception or the prevention of the transmission of sexually transmitted diseases are in Class IIb, unless they are implantable or long term invasive devices, in which case they are in Class III.

Rule 15

All devices intended specifically to be used for disinfecting, cleaning, rinsing or, when appropriate, hydrating contact lenses are in Class IIb. All devices intended specifically to be used for disinfecting medical devices are in Class IIa. Unless they are specifically to be used for disinfecting invasive devices in which case they are in Class IIb.

This rule does not apply to products that are intended to clean medical devices other than contact lenses by means of physical action.

Rule 16

Devices specifically intended for recording of X-ray diagnostic images are in Class IIa.

Rule 17

All devices manufactured utilizing animal tissues or derivatives rendered non-viable are Class III except where such devices are intended to come into contact with intact skin only.

Rule 18

By derogation from other rules, blood bags are in Class IIb

---

**Keep in mind!**

AIMD 90/385/EEC does not contain any classification. All active implants are treated equally. Accessories to active implants are treated like the implants.

---

IVDD 98/78/EC does not contain a definition of classes. Products are differentiated by

- Devices for self-testing

- Devices as listed in Annex II, List A
- Devices as listed in Annex II, list B
- All other products

---

**Exercise:**

Classify some medical devices of your company. In case you are already aware of the risk classes of your products, try to reconstruct the way that has led to the classification.

---

### 3.5. Summary

- Medical devices are classified due to their risk potential in the following classes: I, IIa, IIb and III. Class I can be modified by "s" for sterile products and by "m" for products with a measuring function. The assessment depends on the vulnerability of the human body by the medical device and the place and duration of the application of the medical device. Therefore contact lenses are classified to class IIa, whereas contact lens care products are class IIb.

## Risk Classes

**Application according to the respective directives**

| | High risk | | | Low risk |
|---|---|---|---|---|
| AIMD | All | - | - | |
| MDD | III | IIb | IIa | I |
| IVDD | Annex 2, A | Annex 2, B | Self application | All others |

Fig. 3/1: Overview of Risk Classes for MD, AIMD und IVD

The classification of products and the use of the rules of classification are described in the following guidelines:

- MEDDEV 2.1/3 definition of medical devices vs medicinal products
- MEDDEV 2.1/1 classification of medical devices
- MEDDEV 2.14/1 definition of IVD
- MEDDEV 2.11/1 definition of animal origin
- Consensus statements
- Statements of national authorities (f. e. MHRA)
- www.zlg.de/
- http://ec.europa.eu/enterprise/medical_devices/index_de.htm

---

**Exercise:**

Print out the list of MEDDEV guidelines.

---

Classes or Classifications – What for?

On one hand, there are different procedures to examine and assess the conformity of the products according to the EU directives. On the other hand, different conformity assessment procedures are defined due to different classes/classifications.

3.6. Test Your Knowledge

| | |
|---|---|
| **Q:** | **Which of the following products are medical devices?** |
| | - Surgical suture |
| | - Remedy for a cold (spray) |
| | - Intraocular lenses (IOLs) |
| | - Insulin |
| **A:** | **1 and 3 are medical devices.** |

## 3.7. References

- 93/42/EEC: http://eur-lex.europa.eu/LexUriServ/LexUriServ.do?uri=CONSLEG:1993L0042:20071011:de:PDF

- 90/385/EEC, http://eur-lex.europa.eu/LexUriServ/LexUriServ.do?uri=CONSLEG:1990L0385:20071011:de:PDF

- 98/79/EC, http://eur-lex.europa.eu/LexUriServ/LexUriServ.do?uri=CELEX:31998L0079:en:NOT

## Chapter 4: The Process of CE-Marking
*Dr. Stefan Menzl*

### 4.1. Learning Objective

In this chapter the requirements are described that a product has to meet in order to be able to be distributed in Europe. The EU consists of 27 member states. In order to place a medical device on the European market the manufacturer needs to affix the CE-mark thus stating the compliance with all applicable legal requirements provided by directive 93/42/EEC in its latest revision. The CE-mark must appear in a visible, legible and indelible form on the product directly. If this is not possible, the CE-mark has to be put on the packaging or the accompanying documents.

This chapter describes the requirements that have to be met in order to affix the CE-mark.

### 4.2. Introduction

What do a mobile phone, television-set, intraocular lens and a toy have in common? If you have bought these things in the EU, you will find 2 letters on these products: CE.

The CE-mark is to be found on a lot of products nowadays and both – manufacturers as well as consumers – are aware of it.

For a lot of products the CE-mark is mandatory. It states that the product complies with all legal requirements mandated by applicable directives of the EU, thus guaranteeing protection of health, safety and environmental protection.

Medical devices that have been placed on the market of an EU member state and comply with the local implementation of the applicable European Directive (Medical Devices Act) are also marketable in all other EU member states without further certification.

The medical devices that comply with the European Medical Devices Act bear the CE-mark (CE = Conformité Européenne). Therefore, this CE-mark means that these products can be distributed in the European market and thus, it is a sort of "passport" for these products.

Not every product with a CE-mark is a medical device. Not every medical device has to bear a CE-mark directly. But according to the Medical Device Directive, it is indispensable that the CE-mark is put on the packaging and the instructions for use even if it is not displayed on the product itself. Therefore a CE-mark can be affixed to a television or toy because the manufacturer states the compliance with the requirements. It is not necessary for the CE-mark to appear on the medical device if the latter is too small, its nature does not allow for it or if it is not appropriate, e. g. on an intraocular lens or a coronary stent.

Cave: as far as MDD, AIMDD or IVDD are concerned, only one of these directives can be applied. However, further documents might be relevant or mandatory, e. g. R&TTE (Radio & Telecommunications Terminal Equipment), WEEE (Waste of Electronic and Electrical Equipment) or the Battery Directive (91/157/EEC).

### 4.3. Medical Device – Definition

Medical devices are – according to Article 1 of the Medical Device Directive (93/42/EEC) – "all instruments, apparatus, appliances, software, substances or preparations made from substances or other articles, used alone or in combination, including the software intended by the manufacturer to be used specifically for diagnostic or therapeutic purposes and necessary for the medical device's proper application, intended by the manufacturer to be used for human beings by virtue of their functions, for the purpose of

- Diagnosis, prevention, monitoring, treatment or alleviation of disease
- Diagnosis, monitoring, treatment, alleviation or compensation of injuries or handicaps
- Investigation, replacement or modification of the anatomy or of the physical process
- Control of conception".

An vitro diagnostic used in a laboratory or for self-testing by a lay-person is also considered a medical device. All medical devices are subject to the national Medical Device Act that – together with its regulations – implements the European Directives 90/385/EEC AIMDD, 93/42/EEC MDD and 98/79/EC IVDD.

All medical devices have to comply with the essential requirements and quality assurance measures that are described in detail in these directives thus guaranteeing a high level of protection, performance and safety that means a high quality for patients and users.

The marking is also influenced by the intended use. Only the manufacturer decides for which use his product will be promoted. The same thermometer can be used e. g. for measuring the temperature of a liquid in a laboratory (this thermometer is not a medical device) or for measuring the body temperature of a human being which would classify this thermometer to class I(m) medical device. The intended use (labelling and instruction for use) has to comply with the intended use of the conformity assessment procedure. If this is not the case, there might be regulatory or competitive or even legal consequences.

## 4.4. Significance of CE-marking of a Medical Device

By affixing the CE-mark to his products, a manufacturer states the conformity with the applicable legal requirements defined in the applicable Directive. The CE-mark is the prerequisite to place a product on the market. But there are medical devices that are not allowed to bear the CE-mark: These are samples for clinical investigation and custom-made devices.

Depending on the risk classification of a medical device, a Notified Body has to be involved or not. If a Notified Body is involved, the identification number of that Notified Body has to be added next to the CE-mark.

The compliance with all requirements is proven via a conformity assessment procedure that consists of:

- Safety aspects
  - Analysis, assessment and reduction of risks and adverse events
  - Assessment of the biocompatibility and reduction/elimination of infection risks
  - Electrical safety, electromagnetic safety and mechanical safety
  - Safety in combination with other products
  - Safety and completeness in product labelling and instruction for use
- Efficiency aspects
  - Clinical assessment of medical devices
  - Compliance with all product specifications

    o Proof that product offers the therapeutic or diagnostic benefit that is claimed

    o Guarantee of measurement accuracy

This has to be guaranteed over the whole product life cycle!

## 4.5. Standardization of the CE-marking

In Annex XII of the MDD 93/42/EEC the details concerning the CE-marking of medical devices are described:

"The CE conformity marking shall consist of the initials 'CE' taking the following form:

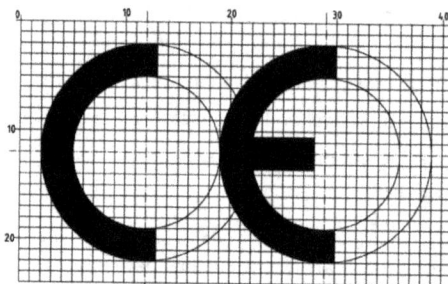

Fig. 4/1: Presentation of the CE mark according to directive 93/42/EWG, Annex XII

- If the marking is reduced or enlarged the proportions given in the above graduated drawing must be respected.

- The various components of the CE marking must have substantially the same vertical dimension, which may not be less than 5 mm.

This minimum dimension may be waived for small-scale devices."

If also other directives apply for a product (e.g. 89/336/EEC or 73/23/EEC) that cover other aspects and also lead to a CE-mark, this has to be mentioned in the accompanying documents.

Fig. 4/2: CE marking of mdc, Stuttgart

If a Notified Body was involved in the conformity assessment, the identification of the NB has to be added to the CE-mark (4-digit code). This only applies to medical devices of class I (sterile or with measuring function), class IIa and IIb and class III. Since no Notified Body is involved in the compliance-assessment of class I devices which are non-sterile and do not have a measuring function, the Notified Body number is not mentioned next to the CE mark. In the above mentioned example, the identification code of the NB "mdc" is added.

The CE-mark is affixed to the product. If this is not possible or appropriate, the CE-mark is put on the package, the instructions for use as well as the sales package.

| **Exercise:** |
| Look for the requirements for custom-made devices in the Annex of the MDD. |

4.6. Regulatory Registration Pathways & Registration Strategies Depending on Product Categories/Classes

The registration strategy is based on the classification of the medical device because it defines whether or not a Notified Body has to be involved.

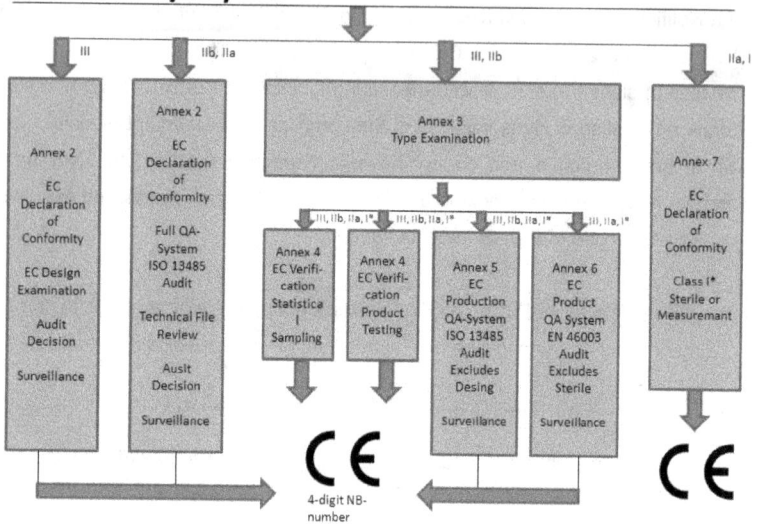

Fig. 4/3: Conformity Assessment Procedures according to 93/42/EWG

A Notified Body has not to be involved for class I-products that are neither sterile nor have a measurement function. Therefore, class I-medical devices only bear the CE-mark.

For all other medical devices, a Notified Body has to be involved in the conformity assessment procedure.

# MDD Class I Routes

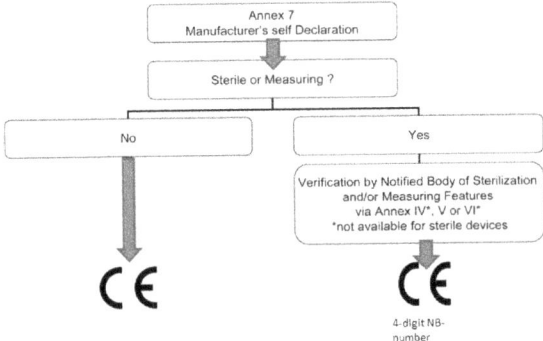

Fig. 4/4: Conformity Assessment Medical Devices Class I

# MDD Class IIa Routes

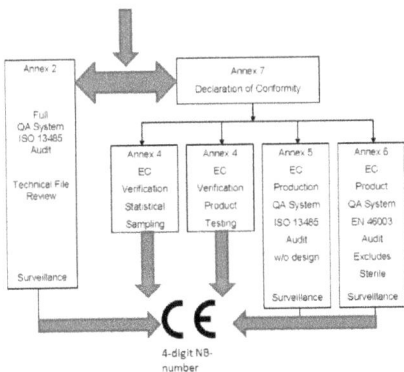

Fig. 4/5: Conformity Assessment Medical Devices Class IIa

# MDD Class IIb Routes

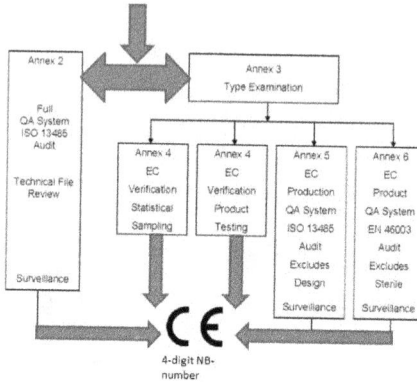

Fig. 4/6: Conformity Assessment Medical Devices Class IIb

# MDD Class III Routes

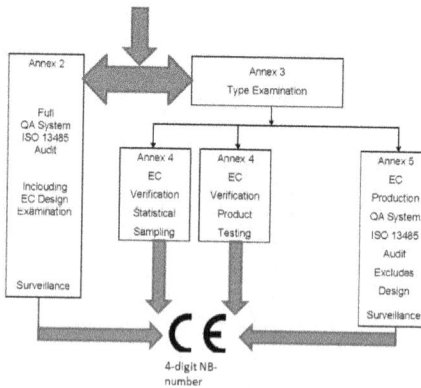

Fig. 4/7: Conformity Assessment Medical Devices Class III

Class I-products only need a self-certification, but a technical documentation has to be established and kept up-to-date (further information in chapter 8).

4.7. Establishing a Regulatory Plan

A regulatory plan for a medical device considers the following elements:

- Directives
- Requirements
- Involving a Notified Body
- Conformity assessment
- Technical documentation
- CE-mark

Directives

First of all a manufacturer has to decide whether Article 1 of the MDD 93/42/EEC applies to the product or whether another directive (e. g. 90/385/EEC or 98/79/EC) is applicable. Consideration also needs to be given to the exclusion criteria defined in Article 1 of MDD 93/42/EEC.

- Directive 93/42/EEC on Medical devices (MDD):
  http://eurlex.europa.eu/LexUri Serv/site/en/consleg/1993/L/ 01993L0042-20031120-en.pdf.
- Further information on 93/42/EEC:
  http://ec.europa.eu/enterprise/policies/single-market-goods/cemarking/professionals/manufacturers/directives/ index_de.htm
- Further information on the Guideline for medical devices:
  http://ec.europa.eu/enterprise/_redirect_template_lang_de.htm

Requirements

The directives of the "New Approach" regarding the CE-mark define a variety of requirements. It is possible that for one product one or more directives apply as well as other regulations.

MDD 93/42/EEC describes the essential requirements a medical device has to meet in order to get the CE-mark. These essential requirements are listed in Annex I of the directive. The compliance with the essential requirements has also to be demonstrated via a clinical assessment according to Annex X of the same directive.

## Involving a Notified Body

Before starting the conformity assessment procedure the responsible manufacturer has to decide whether a self-assessment (medical devices class I) is applicable or whether a Notified Body has to be involved (medical devices class Is, Im, IIa, IIb and III).

In a frequently used CE conformity assessment procedure a Notified Body examines only the quality management system of the manufacturer. As far as medical devices of class IIa, IIb and III are concerned, the design of the product as well as the conformity with the essential requirements is scrutinized by the Notified Body. The Notified Body issues a certificate referring to Annex II to VI of the MDD.

The notified bodies that are accredited for the conformity assessment by the member states are listed in the information system NANDO (New Approach Notified and designated Organizations). On this website, a Notified Body can be searched via the directive, the country or the identification code.

## Conformity Assessment

Depending on the class of medical device, the manufacturer can choose between several (two or more) conformity assessment procedures. In every conformity assessment one or more Annexes (II to VII) of the directive have to be applied. In all cases, a clinical assessment has to be performed and added to the documents that the manufacturer forwards to the Notified Body for assessment. The Notified Body then issues a certificate referring to Annex II to VI of the MDD. In all cases, the manufacturer has to issue a declaration of conformity confirming the compliance with the relevant directive (sole responsibility of manufacturer!). The conformity assessment contains data of the manufacturer (name, address, key specifications of the product), the identification code of the Notified Body (when applicable) as well as a legally binding signature of a responsible person of the organization of the manufacturer.

Technical documentation

Before sending the application to the Notified Body or at latest before the first product is placed on the market, the manufacturer has to establish the technical documentation. This technical documentation must enable an assessment regarding the compliance of the product with the requirements of the directive.

It is the duty of the manufacturer or of his authorized person to keep this documentation at least for 5 years and for implantable devices for at least 15 years after having placed the last product on the European market.

CE-mark

When all above mentioned steps are successfully carried out, a CE-mark has to be affixed to the product. The CE-mark must appear in a visible, legible and indelible form on the product directly. If this is not possible, the CE-mark has to be put on the packaging or the accompanying documents. If also other directives apply for a product that cover other aspects and also lead to a CE-mark, this has to be mentioned in the accompanying documents. If a Notified Body was involved in the conformity assessment, the identification of the NB has to be added to the CE-mark.

4.8. Who is Responsible if a CE-mark is Wrongly Affixed? Who Monitors that Claims are Met?

It is the duty of the manufacturer residing in the EU (or of his authorized person) or of the importer to assure the compliance of the product with the legal requirements. Moreover, for manufacturers based in Germany or Austria, there has to be a safety officer in place, who is also in charge of carrying out post-market surveillance as well as assuring that sales persons are accordingly trained and qualified.

In Germany, the "Medizinprodukte-Sicherheitsplanverordnung" (MPSV, Medical Device Safety Plan Guideline) is the legal basis for this highly responsible task. This regulation defines the consequences for the user and manufacturer. The regulation also describes the procedures of filing, assessing and preventing risks of medical devices that are placed on the market.

The safety officer for medical devices in Germany has to report incidents as well as conducted recalls or safety-relevant measures to the federal higher authority according to § 3 MPSV, § 30:4 MPG). An incident is any malfunction, any failure or deterioration in the characteristics or performance of a medical device as well as any inaccuracy in the labelling or instructions for use which has led or could have led, directly or indirectly, to the death or serious deterioration in the state of health of a patient, user or another person.

The responsible state authorities that monitor the manufacturer, his products as well as professional users are in Germany e. g. Regierungspräsidien (Regional Administrative Councils) or Gewerbeaufsichtsämter (trade supervisory board).

Notified bodies are organizations that audit, certify and verify the assessments of certain medical devices as well as of the quality management system.
The "Zentralstelle der Länder für Gesundheitsschutz bei Arzneimitteln und Medizinprodukten (ZLG, Central Authority of the German Federal States for Health Protection Regarding Medicinal Products and Medical Devices) accredits and monitors the notified bodies.

### 4.9. Additional Marks
Some test-houses and other organizations offer additional markings for medical devices that fulfil certain requirements.
Additional markings for medical devices that already bear the CE-mark result in additional assessments and additional monitoring as well as additional costs.
According to the directive of the European Commission, a medical device can only bear an additional mark in case this mark fulfils another function than that of the CE-mark (e. g. if there is an additional benefit).

### 4.10. Summary
The European Medical Device Directive and its transposition into national law is the basis for product safety because the product has to undergo excessive control measures during the design and manufacturing process. By affixing the CE-mark, the manufacturer confirms the conformity of his product with all applicable legal requirements defined in the European Directive. If a manufacturer wrongly affixed the CE-mark, this is a criminal offence that is penalized.

Depending on the risk classification of a medical device, a Notified Body has to be involved. The identification number of the Notified Body has to be added to the CE-mark.

## 4.11. Test Your Knowledge

| Q1: | **Please classify the following products. Please also state the reason for your classification.** |
| --- | --- |

**Q1: Please classify the following products. Please also state the reason for your classification.**

P1: Sterile gauze bandage

P2: Condom

P3: Laser for grey cataract

P4: ECG electrodes

**A1:**

**P1:** MDD, non-invasive device, contact with injured skin (rule 4), intended to manage the micro-environment of a wound, decide whether special rules apply (rules 13-18), classification: class IIa (rule 4)

**P2:** MDD, invasive, body orifices, not surgical, not implantable or connected to an active medical device, transient use (< 1h), decide whether special rules apply (rules 13-18), classification: class IIb (rule 14)

**P3:** invasive and active product, energy used could lead to harm, special rules not applicable, Classification: class IIb (rule 9)

**P4:** non-invasive, active product, classification without considering EKG monitor, no special rules, Classification: class I (rule 12)

---

**Q2: What significance does the CE-mark have?**

1. The manufacturer states that the product has been manufactured in the EU.

2. Every product is approved by the BfArM.

3. The manufacturer states that his products comply with the applicable European Directive.

4. The manufacturer states that the product is also approved by the FDA.

**A2:** Answer 3 is correct.

| Q3: | When a manufacturer wrongly affixes a CE-mark to his product and places it on the market ... |
|---|---|
| | ... it is no problem |
| | ... it can be penalized as it is a criminal offence |
| | ... this will result in the withdrawal of the certificate |
| | ... the manufacturer has to assess the conformity. |
| A3: | Answer 2 is correct. |

| Q4: | Do custom-made devices bear the CE-mark? |
|---|---|
| A4: | No. |

| Q5: | Do custom-made devices comply with the essential requirements of MDD 93/42/EEC? |
|---|---|
| A5: | Yes. |

| Q6: | Next to the CE-mark there is no identification code. What does that mean? |
|---|---|
| A6: | The medical device is a class I product and is not sterile and does not have a measuring function. |

## 4.12. References

- Richtlinie 93/42/EWG über Medizinprodukte (MDD), http://eur-lex.europa.eu/LexUriServ/site/en/consleg/1993/L/01993L0042-20031120-en.pdf, deutsche Version http://eur-lex.europa.eu/LexUriServ/LexUriServ.do?uri=CELEX:31993L0042:de:HTML

- Directive 93/42/EEC on Medical devices (MDD)

- http://ec.europa.eu/enterprise/policies/single-market-goods/cemarking/professionals/manufacturers/directives/index_de.htm

- http://ec.europa.eu/enterprise/_redirect_template_lang_de.htm

# Chapter 5: Borderline Products

*Dr. Sibylle Scholtz*

## 5.1. Learning Objective

In this chapter you will learn how important a clear demarcation of medical devices is from other products that are meant for healing and alleviation of diseases. This is crucial for the proper implementation of the relevant directives and the correct interpretation and enforcement of applicable national laws. A medical device can be a stand-alone but can be manufactured or used in combination with other substances or products.

The demarcation between medicinal product, cosmetics, biocidal products or a medical device – so-called "borderline products" – is not easy, especially when products or product properties are combined in one product.

## 5.2. Introduction

As the term "borderline" already indicates, these products are in a sensitive border area. Borderline products are considered to be products where it is not clear from the outset whether a given product falls under MDD, AIMDD or the MPD. The classification of these products has to be intensely discussed and assessed. The classification is linked to the *objective* intended use (mode of operation, composition and efficacy) or to the *subjective* intended use (presentation, instructions for use, promotion).

Depending on the classification of a product as medical device, medicinal product or a combination of medicinal product, medical device, biocidal product, cosmetics or commodity, there is a different competent authority responsible that the manufacturer has to contact. This can be the responsible federal authority, the EMA or a notified body. Maybe several authorities have to be involved.

| Product | | Institution / authority |
|---|---|---|
| All medical devices | | Notified Body |
| Medical device and/or medicinal product | | National higher federal authority (e.g. in Germany Bundesinstitut |

| | | für Arzneimittel und Medizinprodukte, BfArM) |
|---|---|---|
| Medical devices of class I | | Federal state authority |
| Medical devices of class II and III | | Notified body resp. federal state authority |
| Biocidial product | | Federal state authority |
| Cosmetics | | Federal state authority |
| Commodities | | Federal state authority |

Fig. 5/1: Which authority for which product?

If the manufacturer has not defined an intended use, a product can be considered to be a commodity, cosmetic or food supplement. But if the product has a medical effect, it has to be differentiated between medicinal product and medical device.

In the EU as well as in the respective member states there are regulations that define the prerequisites for placing products (medicinal products or medical devices) on the market. Borderline products often lead to discussions – on a national and a European level.

For "classical" medical devices that do not contain a pharmacological, immunological or metabolic active substance, a classification to risk classes I to III (MDD 93/42/EEC, Annex X) is applicable. Medical devices that contain a drug are classified according to MDD 93/42/EEC, Annex X, rule 13 as a medical device of risk class III.

The same is true for medical devices incorporating as an integral part a human blood derivative. They are also classified as a medical device of risk class III (MDD 93/42/EEC, Annex X, rule 13).
According to rule 17, all products incorporating tissue of animal origin or animal derivatives are also class III products, unless they are only used on unharmed skin.

Moreover, medical devices can also incorporate blood derivatives, human tissue or human cells or gene therapy medicinal products.

Examples of borderline products:

| Visco-elastic devices | Classification takes place from case to case |
|---|---|
| Eyedrops, used with irritated eyes or general ocular discomfort | Classification takes place from case to case |
| Cremes, which use zinc oxide as ingredient | Classification takes place from case to case |
| Solutions used for disinfection of hands | Biocidial product, no medical device |

Fig. 5/2:  Examples of borderline products

## 5.3. Short Overview of the Underlying Regulations Regarding the Differentiation of Medical Devices and Medicinal Products

There are two directives that were implemented independently of each other: the EU directive 65/65/EEC on the approximation of the provisions laid down by law, regulation or administrative action relating to medicinal products from January 26, 1965 and the MDD 93/42/EEC of June 14, 1993.

E.g. the German legislature defines and regulates medicinal products in the "Arzneimittelgesetz" (AMG, Medicines Act) and medical devices in the "Medizinproduktegesetz" (MPG, Medical Devices Act).

The MDD 93/42/EEC, Annex X, rule 13 clearly defines the differentiation into risk classes for e. g. medical devices with a substance as integral part as follows: "… *the devices incorporating, as an integral part, a substance which, if used separately, can be considered to be a medicinal product, as defined in Article 1 of Directive 2001/83/EC and which is liable to act on the human body with action ancillary to that of the devices, are in class III … .*"

The Arzneimittelgesetz (AMG, Medicines Act) defines medicinal products as "… *substances and preparations of substances that are intended to be applied at or in the human or animal body … indicating the condition, the state or functioning of the body or mental state … .*"

The demarcation between medicinal products and medical devices is stated by the following definition of medicinal product *"... medicinal products are not ... 7. Medical devices and accessories for medical devices according to § 3 of the Medical Devices Act unless it is a medicinal product according to § 2, section 1, no 2 ... ."* This wording defines one exception: physically active IVDs, for example contrast agents that would be a medical device according to the MPG or the MDD, Nevertheless contrast agents according to German law are medicinal products.

Two aspects have to be decided first of all when classifying a product: What is the intended use? What is the main effect of the product? The intended use is defined by the manufacturer and communicated via the product labelling, instructions for use and promotional material. The main effect can be pharmacological, immunological, metabolic, biological or physical. As far as medical devices are concerned, physical and mechanical operating principles are at the forefront (e.g. walking aids, adhesive bandages, implants or prostheses), for medicinal products pharmacological, immunological or metabolic ones.

E. g. the "Medizinproduktegesetz" (MPG, Medical Devices Act) defines medical devices in accordance with the applicable European directives. There is a separate definition for IVDs in the MPG where IVDs are a subgroup of medical devices. The scope of application is differentiated via the following statement *"... is not valid for medicinal products in accordance with § 2, section 1, no 2 of the "Arzneimittelgesetz" (Medicines Act) ... ."* In 2012, the 16th amendment of the AMG (German Medicines Act) more specifically stated that *"... the decision whether a product is a medicinal product or a medical device is mainly driven by considering the main effect of the product ... ."*

The quintessence can be that medical devices and medicinal products shall have a health-related influence on the human being. In contrast to medical devices, medicinal products achieve their main effect via pharmacological, immunological or metabolic mechanisms.

The effects of medical devices on the human body are mainly achieved via physical mechanisms. The term "effectiveness" for medical devices means technical functionality. But there is no definition in a legal sense what the terms "pharmacologically acting products" or "immunologically acting products" or "metabolism" mean and this leaves the field open for excessive discussions and leads to contradicting legal judgements.

The European guideline MEDDEV 2.1 has defined the legal differentiation between medical devices and medicinal products. This guideline is accepted by most courts of European Member-states as state of the art and often serves as an example to support a decision.

The European Union has established an expert group on the differentiation of medical devices that will take uniform decisions on demarcation in the future. First decisions regarding to the demarcation of unanswered questions were published in May 2008.

In Title 2, Article 2 of the directive 2001/83/EC of November 6, 2001 on the community code relating to medicinal products for human use, the legal basis of the application of this directive on borderline products are outlined as follows "… *in cases of doubt, where, taking into account all its characteristics, a product may fall within the definition of a "medicinal product" and within the definition of a product covered by other community legislation, the provisions of this directive shall apply … ."*
That means, in case of doubt a product is classified as a medicinal product!

The reciprocal exclusion of medical devices and personal protection equipment (PPE) was removed. In case that both directives have to be applied, only one CE-mark and one identification code of the notified body can be fixed to the product. Medical devices that also have to be classified via the machinery directive 2006/42/EC have to comply with the essential requirements of the machinery directive that are more specific that the ones in the MDD.

As a rule, a product has to be made subject to a main definition. A product can be either medical device, medicinal product, IVD, cosmetic, food supplement or something else – and therefore only one EU directive applies.
In case there are other supporting mechanisms of action of another product category, additional regulations of that product category directive need to be applied.

### 5.4. Medical Devices that Incorporate a Substance as an Integral Part

In medical technology, medical devices are often combined with substances. This can be for example a drug eluting coronary stent, heparin-coated intraocular lenses or silver-coated wound dressings. A lot of these combinations are distributed in the markets. Combinations of medical devices with a substance are assessed depending on the intended use, the presentation and their mode of action (physical or pharmacological, metabolic or immunological).

The question of classification to medicinal product or medical device focuses on the intended use of the product and the assessment of the main effect. The intended use is not only proved via the technical documentation but also via the instructions for use and the promotional material of the product. The main effect can be described with "pharmacological, immunological and metabolic" (= medicinal product) or "physical and/or mechanical" (= medical device).

Two documents are relevant to medicinal products:
- Directive 65/65/EC, 2001/83/EC amended by 2004/27/EC
- National Medicines Act and related legal regulations

In order to classify a medical device, the following MEDDEV guidelines are helpful:
- MEDDEV 2.1/1: Definitions of medical devices, accessory and manufacturer
- MEDDEV 2.1/3, 12/2009: Borderline products, drug delivery products and medical devices incorporating, as an integral part, an ancillary medicinal substance or an ancillary human blood derivative
- MEDDEV 2.4/1: Classification of medical devices

---

**Exercise:**

Look up all the above mentioned legal texts and documents in the www.

---

For combinations of medical devices and medicinal products a special rule applies. These combination products can only be classified to one category. The classification cannot remain open because this would result in substantial consequences for distribution including promotion and reimbursement by the health insurance funds.

The classification as medicinal product has also consequences for the distribution because an approval for a medicinal product is only granted for defined EU member states whereas medical devices can be distributed in all EU member states if a conformity assessment procedure has been conducted, a CE-mark affixed and if the product labelling complies with the national language requirements.

Also, different requirements and notification obligations apply regarding vigilance. When promoting a product one has to ensure that the promotional product messages are also relevant to the intended use of a medical device. The manufacturer has to describe the therapeutic auxiliary function with great care.

Regulations regarding advertising of medicinal products or medical devices differ significantly per EU member state.

Medicinal products are classified into two groups:

**Medicinal product by presentation:** the classification depends on the indication and/or appearance. The ingredients are not taken into consideration.

**Medicinal product by function:** The pharmacological ingredients are of high relevance, the appearance is not.

Where a device incorporates, as an integral part, a substance which, if used separately, may be considered to be a medicinal product as defined in Article 1 of Directive 2001/83/EC and which is liable to act upon the body with action ancillary to that of the device, the quality, safety and usefulness of the substance must be verified by analogy with the methods specified in Annex I of Directive 2001/83/EC" (MDD, Annex I, 7.4)."

The classification of a medical device takes place according to Annex IX, rule 13 of the MDD 93/42/EEC.

According to MEDDEV 2.1/3 (list B.4.1.) the following examples of combinations of medicinal product and medical device are to be classified as medical devices:

- Drug eluting coronary stents
- Catheters coated with heparin or an antibiotic agent
- Intrauterine contraceptives containing copper or silver
- Condoms coated with spermicides
- Soft tissue fillers incorporating local anaesthetics

- Bone cements containing antibiotic
- Root canal fillers incorporating local anaesthetics
- Electrodes with steroid-coated tip
- Wound dressings, surgical or barrier drapes (including tulle dressings) with antimicrobial agent

The EU defines 6 possible combinations of medicinal products and medical devices:
- Drug-delivery medical device
- Medical device and medicinal product one product for single use
- Medical device, incorporating, as an integral part, an ancillary medicinal substance (class III, medical device)
- Medical device, incorporating, as integral part, an ancillary human blood derivative
- Combination of advanced therapy medicinal product and medical device
- IVD in combination with medicinal product

Various variants of combinations of medicinal products and medical devices are possible. In order to decide to which group the product is classified, the main effect of its part has to be taken into consideration:
- A combination of a patch (medical device) with a medicinal product = transdermal patch (e. g. Nicotine patch, hormone patch). This is a medicinal product.
- If medical device and medicinal product are manufactured separately but are inextricably dependent on this combination, e.g. prefilled syringes or asthma spray applicator, the product is a medicinal device (administration device has to be considered).
- The medicinal product and medical device form a packaging unit but both products are separately available, e. g. a cough syrup and a measuring spoon. The measuring spoon is a medical device of class Im (class I with measurement function), the cough syrup is a medicinal product. Both parts of this set have to undergo the appropriate procedure.
- If it is a medical device with an ancillary medicinal product, e.g. wound dressings with an antimicrobial agent, bone cement containing antibiotic or

intraocular lenses coated with heparin a MD conformity assessment procedure with a consultation procedure has to be conducted.

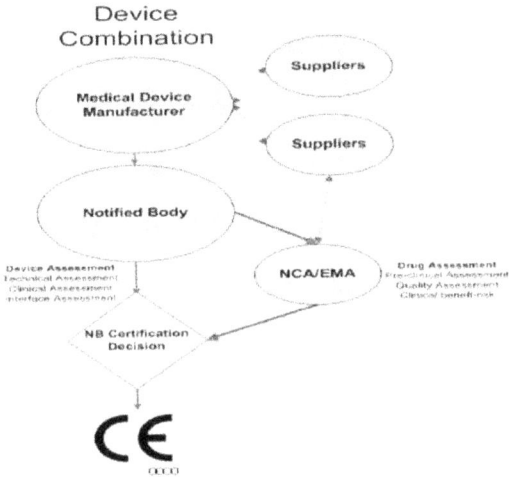

Fig. 5/3: Overview over the synergy of a conformity assessment at consultation procedure

---

**Exercise:**

**Please classify the following products:**

Condom

- Pregnancy test

- Artificial tears

- Alcohol

- Sterilizer

- Dialysis solution

- Drug-coated stents

**Answers:**

- Condom = medical device

- Pregnancy test = IVD

- Artificial tears = both is possible, medical device or medicinal product

---

---

- Alcohol = depending on the application:
    - for disinfection of skin = medical device
    - for disinfection of surfaces = biocidal product
    - in combination with a medical device = medical device
- Sterilizer = when used for the sterilization of medical devices = medical device or an accessory to a medical device
- Dialysis solution = medical device, except when used for peritoneal dialysis (= medicinal product)
- Drug-coated stents = medical device

---

**Consultation Procedure**

Products that are a combination of a medical device and a medicinal product are subject to e.g. the Medizinproduktegesetz (Medical Devices Act) when the main effect is NOT achieved via pharmacological, immunological or metabolic pathway. But the drug component is assessed according to the EU directives. The complete product will be assessed according to the MPG by a notified body. In order to assess the requirements of the drug component, the notified body consults a competent authority for medicines of an EU member state of its choice. This consultation procedure serves to ascertain whether the medical device and the medicinal product are compatible with each other. The manufacturer should contact the competent authority for medicines early – before submitting the request for consultation. Procedural questions and critical issues (f. e. the effect of the drug component) can be discussed in a meeting. The better prepared, the shorter the proceedings duration time (about 6 months). The manufacturer should also closely cooperate with the notified body that will submit the application.

The consultation is the time-critical part of a conformity assessment whereas the clinical assessment is the time-critical part of the CE certification procedure (source: BVMed, www.bvmed.de/themen/medizinprodukteindustrie-1/CE-Kennezichnung/pressmitteilung/Konsultationsverfahren_fuer_Medizinprodukte,_die_Arzeimittel_enthalten,_sind_noch_nicht_etabliert.html).

The competent authorities for medicines are listed on the website www.gmp-navigator.com/nav_link_behoerden.html.

The purpose of the consultation is described in MEDDEV 2.1/3, C1. On one hand, the medicinal substance that is incorporated in a medical device should be verified according to its quality, safety and clinical risk-benefit-ratio, taking into account the intended use of the medical device. On the other hand, a consultation of the notified body on the risks of the substance is ensured that are already known to the competent authority for medicines.

As far as combination products are concerned, a thorough preparation is key. It would be a good idea to choose a notified body that is experienced in combination products.
The notified body has a key role in …

- Assessing the registration strategy as well as the strategy for the clinical assessment (optional)
- Assessment of the technical documentation (design dossier), audit of the manufacturer or critical suppliers
- Obtaining of an assessment of a drug component at the competent authority
- Final assessment and issuing of a CE certificate

Parts of a consultation documentation:
Usually, incorporated drugs like antibiotics or heparin have the same intended use like the already approved drug. Therefore the consultation procedure is not about the approval of the drug itself. If there is a new intended use of a substance, this has to be assessed and documented according to currently applicable directives on pharmaceutical products. The result is not an approval but a statement. The same is true for new substances. The consultation file is put together by the notified body. It is recommended to establish the design dossier and a second dossier in form of a consultation file and submit both to the notified body.
See MEDDEV 2.1/3 for the structure. At the beginning of the consultation documentation there is the description of the medical device that has to contain a justification why the drug was incorporated in the device and the benefit provided by the substance. Moreover, the results of the risk analysis as well as the critical assessment of the single risks have to be added.

The notified body is the applicant in the consultation procedure at the medicines authority. The authority issues a statement, the decision of issuing a CE certificate is up to the notified body. There is no legal definition for the duration of the procedure. Consultation procedures are high priority procedures and are usually finalised within 6 months. The requirements for the content of the documentation are that they have to be according to the current scientific level of knowledge, proof of the quality of the drug, of its safety and its benefit.

Notified Bodies have frequently complaint about the fact that often parts of the documents to be submitted are missing as well as the justification of the classification as medical device with a substance as integral part or the assessment by the notified body. Often the main effect of the substance is not known, studies to the substance are not available. Sometimes the documentation is not state-of-the-art or there is no profound benefit-risk assessment.

From the viewpoint of the notified body, manufacturers of medical devices with substances as an integral part should contact their notified body as soon as possible. In case there are differences of opinion regarding the consultation procedure between the manufacturer and the notified body (NB), the NB has to submit the case to the competent authority for decision. The experience with this procedure shows that this takes about 6 to 9 months. 24 months are also possible in specific cases. Also the quality of the assessment is quite different. As far as the fees are concerned, there are no clear definitions, for example by the EMEA and the German BfArM (The Federal Institute for Drugs and Medical Devices).
As far as the assessment duration is concerned there are only clearly defined period of time by the EMEA and to a certain extent by the MEB in the Netherlands. If the assessment by the Authority for Medicines results in a negative expert opinion, the product cannot be certified. Negotiations with the competent authority at an early stage are possible and this option should also be seized.

It is important to have discussions at an early stage with the notified body in order to establish a set of measures. The classification has to be defined and whether it is a known substance or whether additional clinical studies have to be conducted. Moreover, it is important to choose a competent authority experienced with consultation procedures as well as to have early discussion with the chosen authority regarding the classification of the product. The notified body has an advisory role when choosing a competent authority. The manufacturer has to provide resources for answering questions about the dossier: questions by the notified body regarding the MD part and questions by the authority regarding the substance (as integral part).

Example: Drug-Coated Stents (Medical Device)

Medical devices as well as medicinal products undergo different mandatory quality assurance phases.

Medical devices are assessed – after technical development and production – according to biocompatibility, technical performance and clinical assessment before the CE certificate is granted. After receiving the CE certificate, the manufacturer has to ensure the long-term surveillance.

Substances undergo – after the technical development and production – tests regarding toxicology, pharmacodynamics and pharmacokinetics, application on test subjects and patients (phase I to III) before they are approved. After approval they also have to be monitored on a long-term basis (phase IV studies).

For a stent that releases a new substance, the key components of this product are the stent, the catheter application system, the polymer coating as well as the drug substance. The cooperation with the provider of the drug substance requires an intensive project management. The clinical assessment of non-approved substances can only be achieved via clinical trials. The consultation procedure of drugs produced via genetic engineering or via human blood or plasma has to be conducted by the EMEA. For this project a design assessment of at least 12 months is advisable. The consultation process is not a common and well established process in Europe at the moment. This is especially true for medical devices with non-approved substances (source: www.bvmed.de/themen/medizinprodukteindustrie-1/CE-Kennzeichnung/pressemitteilung/Konsultationsverfahren_fuer_Medizinprodukte,_die_Arzneimittel_enthalten,_sind_noch_nicht_etbaliert.html).

The German BfArM (Federal Institute for Drugs and Medical Devices) has also published general information on consultation procedures (www.bfarm.de/DE/Arzneimittel/2_zulassung/zu/Arten/natVerf/HinweiseKonsultationsverfahren.html?nn=1012262):

"... *information on conducting consultation procedures and submission of documents for medical devices with substances as an integral part ...*" (March 4, 2008).

Products that are a combination of medical device and medicinal product (e.g. catheter with heparin coating) are subject to the Medical Devices Act in case that their main effect is not achieved via pharmacological, immunological or metabolic action. The medicinal component, however, has to be assessed according to directive 2001/83/EC (formerly: 75/318/EEC) and 2003/63/EC and the referring guidelines. The final product will be assessed by the notified body according to the Medical Devices Act. As far as the medicinal component is concerned, the notified body consults a national approval authority (which is in Germany the BfArM). This authority assesses according to MDD 93/42/EEC, Annex I, 7.4 the safety, quality and benefit of the substance while taking into consideration the intended use of the medical device.

If the substance incorporated in the medical device contains derivatives of human blood or blood plasma, the EMEA has to be consulted (see directive 2000/70/EC).

The notified body has to take into consideration the expert opinion of the approval authority that resulted from the consultation procedure. Because this expert opinion is the final decision of this authority (see MDD 93/42/EEC, Annex II, 4.3 and Annex III, 5).

Intensive discussions with the BfArM at an early stage are possible. Especially if there are complex questions it is advised to contact the BfArM as early as possible. Fees may be charged according to the fees regulation of the Medical Devices Act and the derived legal regulations (March 27, 2002 that was revised February 16, 2007)."

The BfArM also published a proposal to the course of the consultation procedure (www.bfarm.de/DE/Arzneimittel/2_zulassung/zuArten/natVerf/ablauf-konsult_verfahren.html?nn=101069):

"... *Course of a consultation procedure at the BfArM* (March 22, 2004)

Consultation procedures for the assessment of a substance being an integral part of a medical device is of high priority for the BfArM. In case a notified body plans to ask the BfArM to assess such a substance, the NB contacts the BfArM in order to negotiate the time of submission. The communication therefore takes place between BfArM and notified body. The NB communicates with the MD manufacturer.

1. The NB first of all applies for a submission code via fax (fax no: +49-228-207-3681) adding the following information: name of the manufacturer, name of the NB, labelling of the medical device, classification of the medical device and of the substance in advance for the MD to be assessed. The NB receives the submission code via fax.

2. The NB submits the complete documentation mentioning the submission code at the BfArM and receives an acknowledgement of receipt.

3. The BfArM first of all assesses the classification of the product (medicinal product, medical device or medical device with a substance as an integral part), if the product is within the fields of competence of the authority, the familiarity with the substance according to the well established medicinal use of the directive 2003/63/EC and the notice to applicants or the classification as a new substance according to AMG §48, section 2, 1 (because the effects and side effects of the substance in not known and not assignable in the medical science). Moreover, the completeness of the documentation is assessed. If the documentation is not complete, the notified body is asked to submit the missing documents within one month. The documentation is then distributed internally to the respective specialized divisions that assess the content of the documentation. After this assessment the BfArM issues an expert opinion on the quality, effect and safety of the substance. In case that shortcomings have been identified, the notified body together with the manufacturer gets the chance to rectify these shortcomings within a defined period of time (two-phase procedure).

4. The expert opinion of the quality, benefit, usefulness and safety of the substance are sent to the notified body in a final statement.

5. The notified body then informs the BfArM on further proceedings (certification, new consultation, a switch to another approval authority after negative expert opinion). ..."

The 15-page application form for the consultation procedure is published e. g. on the website of the BfArM (www.bfarm.de/DE/Arzneimittel/2_zulassung/zulArten/natVerf/_form/functions/forma mzul-node.html).

### 5.5. Medical Devices Incorporating Material of Animal Origin, Human Blood, Human Plasma or Human Tissue

For the above mentioned medical devices the following regulatory texts have to be taken into consideration:

- MDD 93/42/EEC, Annex IX, rule 13
- Manual on Borderline and Classification in the Community Regulatory Framework for Medical Devices
- National Medical Devices Act and its executing legal regulations

MDD 93/42/EEC, Annex IX, rule 13, 4.1 clearly defines that "all devices incorporating, as an integral part, a human blood derivative are in Class III."

The MDD cannot be applied on tissue transplants, on cells of human origin or on products that incorporate parts or cells of human origin. The exception to this definition are products that are medical devices incorporating, as an integral part, a substance which, if used separately, can be considered to be a medicinal product or a device incorporating, as an integral part, a human blood or human plasma derivative (Directive 2001/83/EC) and that have an ancillary action to that of the medical device.

The same is true for transplants or tissue or cells of animal origin – as long as they are not used as part (containing components of animal non-viable tissue or being derived from it) of the medical device.

Gene therapy, cell therapy or tissue-generated products (ATMP: Advanced Therapy Medicinal Products), can be combined with medical devices, for example as frame for tissue formation, implants, stents or circulation outside the body.

In case of combination medical devices with ATMP, the assessment of the ATMP component is performed centrally by the Committee for Advanced Therapies (CAT) of the EMEA. In this case, the CAT contacts the notified body in order to get an expertise on the medical device by the NB.

5.6. Summary

Medical devices can be combined with substances, IVDs, ATMP products or material of animal origin or with products derived from human blood or plasma.

Key to a successful registration strategy is the correct classification of the product. In order to classify a product the intended use (defined by the manufacturer) and the main effect of the product are key. In case the product is effective via physical mechanisms it is a medical device. If the main effect of the product is achieved mainly via pharmacological, immunological or metabolic action the product is considered to be a medicinal product.

Principally, a product has only be classified to one category that means it is only subject to one European directive (medical device directive, medicinal product directive or IVD directive). However, certain individual relevant aspects of other directives or regulation might apply.

But beside the main effect, there also can be auxiliary effects. In this case, assessment criteria of the product category are of importance, from which this effect derives. A classic example of this is a combination of a medical device and a substance. The product is then subject to the MDD because the main effect is due to physical mechanisms). Auxiliary effects via the substance component (e. g. heparin coating) are present.

The MDD defines the classification of the product (class III) and the applicable so-called consultation procedure that is described in detail in the MEDDEV document.

Responsible for the assessment of compliance with the essential requirements is the notified body that has to come to a mutual agreement with the medicines authority via a consultation procedure. The medicines authority assesses the substance component especially considering its application in combination with the medical device. If this results in a positive assessment, the notified body finalises the conformity assessment by issuing a CE certificate.

Similar procedures exist for the combination of medical devices with material of animal origin as integral part or IVDs with a substance as integral part.

In case of ATMP combination products the responsibilities are as follows: there is a central assessment of the ATMP component by the Committee for Advanced Therapies (CAT) of the EMEA. In this case the CAT contacts the notified body in order to get an expertise on the medical device by the NB.

### 5.7. Helpful Documents

- CONSOLIDATED Active Implantable Medical Devices Directive (90/385/EEC)
- CONSOLIDATED Cosmetic Products Directive (76/768/EEC)
- CONSOLIDATED Medical Device Directive (93/42/EEC)
- In-Vitro Diagnostic Devices Directive (98/79/EC)
- Personal Protective Equipment Directive (89/686/EEC
- Borderline and Classification issues page of the EC website:
  http://ec.europa.eu/health/medical-
  devices/documents/borderline/index_en.htm

Guidance Documents

- GUIDANCE Document on Interpreting Directive 2007/47/EC
- MEDDEV 2.1/1 - Defines Medical Devices, Accessories and Manufacturer

- MEDDEV 2.1/3 - Demarcation between MDD and Medicinal Products Directive
- MEDDEV 2.1/4 - Discusses demarcation between the EMC and PPE Directives
- MEDDEV 2.14/1 - Borderline issues between the IVD and Medical Device Directives
- MEDDEV 2.4/1 - Classification of medical devices
- MEDDEV 2.5-8 - Deals with assessment of devices with animal materials
- Manual on Borderline and Classification in the Community Regulatory Framework for Medical Devices
  http://ec.europa.eu/health/medical-
  devices/files/wg_minutes_member_lists/borderline_manual_ol_en.pdf

- MEDDEV on IVD Borderline Issues

  http://ec.europa.eu/consumers/sectors/medical-
  devices/files/meddev/2_14_ivd_borderline_issues_jan2004_en.pdf

- Biocides Manual

  http://ec.europa.eu/environment/biocides/pdf/mod_040705.pdf

- IMB AND MHRA guidance http://www.imb.ie/EN/Medical-Devices.aspx,

  http://www.mhra.gov.uk/Publications/Regulatoryguidance/Devices/index.htm

## 5.8. Test Your Knowledge

**Q1:** **For which products is a consultation procedure necessary in connection with a conformity assessment?**

1. Medical devices of class I

2. Medical devices incorporating a pharmacological substance

3. Medical devices that incorporate, as integral part, a human blood derivative

4. This is only necessary for medicinal products

**A1:** The right answer is 3.

**Q2:** **Who conducts a consultation procedure?**

1. the manufacturer

2. the BfArM

3. The e. g. Regierungspräsidium (Regional Administrative Council)

4. The notified body in cooperation with an authority for Medicines of the EU member state

**A2:** Answer 4 is correct.

## 5.9. References

- MDD 93/42/EC Annex IX
- German Medicine Act (AMG)
- 2001/83/EC
- 65/65/EG, 2001/83/EG, revised version: 2004/27/EG
- MEDDEV 2.1/1: Definitions of "medical devices", "accessory" and "manufacturer"

- MEDDEV 2.1/3, 12/2009: Borderline products, drug delivery products and medical devices incorporating, as an integral part, an ancillary medicinal substance or an ancillary human blood derivate
- MEDDEV 2.4/1: Classification of medical devices
- Information on the consultation procedure by the BVMed www.bvmed.de/themen/medizinprodukteindustrie-1/CE-Kennzeichnung/pressemitteilung/Konsultationsverfahren_fuer_Medizinprodukt e,_die_Arzneimittel_enthalten,_sind_noch_nicht_etabliert.html
- Overview of authorities for medicines: http://www.gmp-navigator.com/nav_link_behoerden.html
- Information on the consultation procedure by the BfArM*: www.bfarm.de/DE/Arzneimittel/2_zulassung/zulArten/natVerf/HinweiseKonsult ationsverfahren.html?nn=1012262):
- Course of the consultation procedure, BfArM: www.bfarm.de/DE/Arzneimittel/2_zulassung/zulArten/natVerf/ablauf_konsult_v erfahren.html?nn=101069
- Application of the consultation procedure, BfArM: www.bfarm.de/DE/Arzneimittel/2_zulassung/zulArten/natVerf/form/functions/fo rmamzul-node.html
- Also see chapter 5.7 on helpful documents
- BfArM: Federal Institue for Medicines and Medical Devices

## Chapter 6: Software as Medical Device

*Dr. Sibylle Scholtz*

### 6.1. Learning Objective

In this chapter you will learn that – according to the Medial Devices Act and international directives – software can be an independent medical device as well as a part of another medical device. Therefore, software is subject to regulatory requirements that result in requirements for the QM system of a manufacturer.

### 6.2. Introduction

Modern medical appliances that are classified as medical devices (e. g. medical lasers, phaco machines, X-ray device, CT-machines) need software in order to operate. These programs define the functionality of the devices. The software is an integral part of the device without which the device would not work. That's why this software is also subject to regulatory processes and has to meet the legal requirements like any other medical device. This is clearly defined in the Directive 2007/47/EC that is described in the following part.

1. All three directives for medical devices also refer to the use of software: Directive 93/42/EEC from June 14, 1993 on medical devices: Article 1 "... *'medical device' means any instrument, apparatus, appliance, software, material or other article, whether used alone or in combination, including the **software** intended by its manufacturer to be used specifically for diagnostic and/or therapeutic purposes and necessary for its proper application, intended by the manufacturer to be used for human beings for the purpose of ...*" In Annex I, part II, 12.1a it is stated that " *... for devices which incorporate **software** or which are medical **software** in themselves, the **software** must be validated according to the state of the art taking into account the principles of development lifecycle, risk management, validation and verification ... .*"

2. In Directive 98/79/EC from October 27, 1998 on in vitro diagnostic (IVD) medical devices, it can be read that " *... (8) Whereas instruments, apparatus, appliances, materials or other articles, including software, which are intended to be used for research purposes, without any medical objective, are not regarded as devices for performance evaluation ... .*" In Annex I, part B, 6.1 the following description can be found: "*... Devices incorporating electronic*

*programmable systems, including* **software**, *must be designed to ensure the repeatability, reliability and performance of these systems according to the intended use ... ."*

3. Directive 90/385/EEC relating to active implantable medical devices:

Article 1 (2) *"... For the purposes of this Directive, the following definitions shall apply:*

*(a) 'medical device' means any instrument, apparatus, appliance, software, material or other article, whether used alone or in combination, together with any accessories, including the software intended by its manufacturer to be used specifically for diagnostic and/or therapeutic purposes and necessary for its proper application, intended by the manufacturer to be used for human beings for the purpose of ... ."*

In Annex I, part II, 9 follows: *"... The devices must be designed and manufactured in such a way as to guarantee the characteristics and performances referred to in I. 'General requirements', with particular attention being paid to proper functioning of the programming and control systems, including software. For devices which incorporate software or which are medical software in themselves, the software must be validated according to the state of the art taking into account the principles of development lifecycle, risk management, validation and verification ... ."*

All three directives define the software lifecycle, a validation due to the state of the art technical development, repeatability precision and the risk management as important parts of the essential requirements. First of all one has to decide, whether the software is an independent medical device or is used in combination with a medical device.

6.3. Classification

As already shown, software is covered by several EU directives. Which directive applies depends on the <u>intended use</u> of the medical device. If software is a medical device, it has to have a medical intended use as a medical device, an active implantable device or an in vitro diagnostic device. Depending on the intended use the conformity assessment procedure is chosen. The software has to comply with the essential requirements and the harmonized standards that should be reflected in the QM system.

The following harmonized standards refer to software:

- EN 62304: Medical Device Software, Software Life Cycle Processes
- EN ISO 14971: Medical Devices – Application of Risk Management to Medical Devices
- EN 60601-1: Medical Electric Equipment: Part 1: General requirements for Basic Safety and Essential Performance
- EN 60601-1-4: Medical Electric Equipment: General Requirements for Safety, Collateral Standard: Programmable Electrical Medical Systems
- EN 60601-1-6: Medical Electric Equipment: General requirements for Basic Safety and Essential Performance, Collateral Standard: Usability
- EN 62366: Medical Devices: Application of Usability Engineering to Medical Devices
- EN 62083: Medical Electric Equipment: Requirements for the Safety of Radiotherapy Treatment Planning Systems

Especially for software that is used with IVDs the following list applies:

- ISO 13485: Quality Management Standard for Medical Devices and Related Services
- ISO 14971: Medical Devices: Application of Risk Management to Medical Devices
- EN 62304: Medical Devices Software: Software Life-cycle Processes
- IEC 62366: Medical Devices: Application of Usability Engineering to Medical Devices
- EN 591: Instructions for Use for IVD Instruments for professionals
- EN 592: Instructions for Use for IVD Instruments for Self-testing
- EN 375: Information Supplied by the Manufacturer with IVD Reagents for Professional Use
- EN 376: Information Supplied by the Manufacturer with IVD Reagents for Self-testing

- EN ISO 18113-1: IVD Medical Devices: Information Supplied by the Manufacturer (Labelling) Part 1: Terms, Definitions and General Requirements, Part 3: IVD Instruments for Professional Use
- EN 1041: Information Supplied by the Manufacturer for Medical Devices
- EN 980: Requirements for Symbols Used with Medical Devices
- ISO 15223: Requirements for Symbols Used in Medical Devices Labelling that Convey Information on the Safe and Effective Use
- EN 15225: Medical Devices Nomenclature Data Structure
- EN ISO 17511: IVD Medical Devices: Measurement of Quantities in Biological Samples – Metrological Traceability of Values Assigned to Calibrators and Control Material
- EN ISO 18153: IVD Medical Devices: Measurement of Quantities in Biological Samples – Metrological Traceability of Values for Catalytic Concentration of Enzymes Assigned Calibrators and Control Material
- ISO 14155: Clinical Investigation of Medical Devices for Human Subjects – Good Clinical Practice, Part 2: Clinical Investigation Plans
- MEDDEV 2.14/3, MEDDEV 2.5/5, MEDDEV 2.12/1 National laws

### 6.4. Requirements to the QM System of the Manufacturer for Software

In EN 62304 the regulatory requirement to software that should be classified as a medical device is described in the following chapters:

- Chapter 5: Software Development Process
- Chapter 6: Software Maintenance Process
- Chapter 7: Risk Management Process
- Chapter 8: Configuration Management Process
- Chapter 9: Problem Resolution Process

After that the risk management activities as defined by ISO 14971 and the general quality management should be considered.

### 6.5. Conformity Assessment

The conformity assessment procedure should clearly demonstrate that the manufacturer has thoroughly planned, implemented and verified every step of the development of the product (and complied with the requirements of every process step).

In the **design process** the following has to be scrutinized and documented: the definition of the product requirements, the product verification and product validation, references to standards, design methods, tools, traceability, change and configuration management as well as historical steps. The software algorithm is "established" in the design phase.

In the **manufacturing process** the following aspects are indispensable: process validation and process monitoring, error management and traceability. "Manufacturing" is the process of duplicating and saving information on CDs, DVDs or USBs as well as downloads from the worldwide web. Another essential requirement is the labelling of the packaging. In this process step, the choice of appliances takes place and the validation of used software tools or processes as well as the documentation of traceability.

The standards that should be applied in the design and manufacturing process are listed below:

| Process | EN 13485 | EN 63204 | EN 14971 | EN 13612 (IVD) |
|---|---|---|---|---|
| Design and Development | x | x | | |
| Product requirements | x | x | | |
| Development of Software Design | | x | | |
| Validation | x | | x | x |
| Verification | x | x | x | |
| Change | x | x | x | x |
| Configuration | | x | | |
| Traceability | x | x | | |
| Risk Management | | x | x | |

| | | | | |
|---|---|---|---|---|
| Correction / Prevention | X | | | |
| Process Validation | X | | | X |
| Process Monitoring | X | | | |
| Error Management | X | | | |

Table 6/1: Overview over standards/norms relevant for design and manufacturing

When placing software on the market, the sales department has to guarantee that the mark is visibly affixed and that the configuration management, traceability and process validation is working.

Especially as far as installation and service is concerned, the specific product requirements and the reliable traceability have to be met.

At any time of the above mentioned process steps the following has to be reliably implemented: risk management, corrective and preventive actions as well as reporting/vigilance.

| Process | EN 13485 | EN 63204 | EN 14971 | MEDDEV 2.12-1 |
|---|---|---|---|---|
| Traceability | X | | | X |
| Risk management | | X | X | X |
| Reporting to relevant Authorities (vigilance) | | X | | X |
| CAPA (corrective and preventive actions) | X | X | | X |
| Configuration management | | X | | X |
| Process validation | X | | | |

Table 6/2: Overview over standards/norms relevant for sales, installation and maintenance

In the MEDDEV guideline 2.1/6 from January 2012 the following overview is published that is helpful in the assessment of software as medical device.

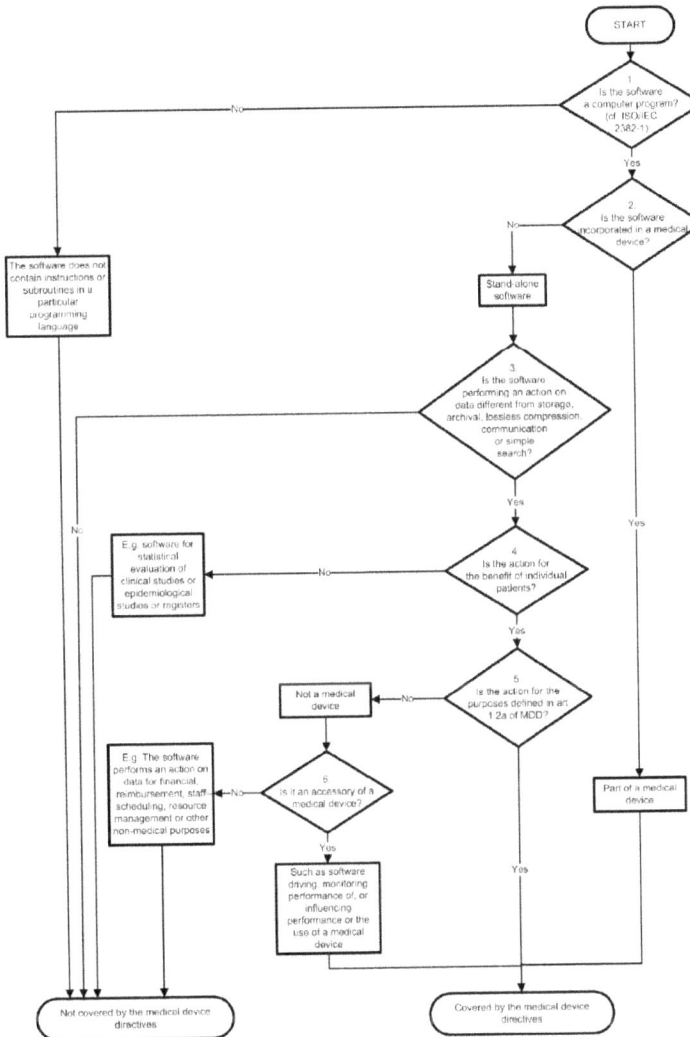

**Fig. 6/1: Decision tree to assist qualification of software as medical device**

(http://ec.europa.eu/health/medical-devices/files/meddev/2_1_6_ol_en.pdf)

6.6. Summary

The cooperation with experts regarding regulations and standards – when assessing software as medical devices – can be very helpful because these experts are well acquainted with the content of these regulations and standards and therefore can interpret them well. Also joining committees like committees of DIN, VDE or EUROCAT BSI can be helpful.

6.7. Test Your Knowledge

| | |
|---|---|
| **Q1:** | When a manufacturer wants to place a medical device software of risk class I on the market, the essential requirements of the MDD 2007/47/EC have to be met. In order to comply with 12.1a of this directive, the manufacturer can e. g. use the harmonized standard DIN EN 62304 (Medical Software). This standard asks as a reference for risk management DIN EN 14971 and for quality management systems DIN EN 13485. Does that mean that the manufacturer of such software should have or has to have a QM system according to DIN EN 13485? |
| **A1:** | The compliance with standards (also harmonized standards) is voluntary. Using a harmonized standard results in the presumption of conformity. A reference in a standard to another standard from my point of view does not lead to an obligation. A manufacturer who does not meet ISO 13485, may not be able to claim full compliance with this standard. But this full compliance is not conclusively demanded, therefore this should not be a problem because the manufacturer can refer to an accepted quality management system. |

| | |
|---|---|
| **Q2:** | A manufacturer developed a software that documents the measurement results of the visual faculty of patients before a laser treatment as well as afterwards on a yearly basis. During the process, the data is transferred directly from the refractometer. Is this software a medical device? Do the essential requirements according to MDD have to be met and documented? |
| **A2:** | No. This software only documents the results. An alleviation of pain or a treatment does not take place. As it is not a medical device the requirements of MDD do not apply. |

## 6.8. References

- 93/42/EWG http://eur-lex.europa.eu/LexUriServ/LexUriServ.do?uri=CONSLEG:1993L0042:20071011:de:PDF

- 90/385/EWG, http://eur-lex.europa.eu/LexUriServ/LexUriServ.do?uri=CONSLEG:1990L0385:20071011:de:PDF

- 98/79/EC, http://eur-lex.europa.eu/LexUriServ/LexUriServ.do?uri=CELEX:31998L0079:en:NOT

**Chapter 7: Essential Requirements**

*Dr. Stefan Menzl*

7.1. Learning Objective

In this chapter you will learn about the importance of the essential requirements defined by the EU Medical Device Directive, about the technical documentation and the CE marking process. You will understand the consequences derived from these requirements to the product documentation and will be able to implement them accordingly.

7.2. Introduction

In the "Old Approach" every country had its own requirements for placing medical devices on the market. The prerequisites for the registration were to a certain degree very different. Manufacturers had to register a product in every country, product-specific standards were often binding nationally. This resulted in different requirements to the design of the products.

In the "NEW Approach" the essential requirements for medical devices are described in the Annex I of the relevant EU directives:

- In the Annex I of the active Implantable Medical Device Directive 90/385/EEC
- In the Annex I of the In Vitro Diagnostic Medical Device Directive 98/79/EEC
- In the Annex I of the Medical Device Directive 93/42/EEC

Complying with these essential requirements is legally mandatory and is the responsibility of the manufacturer (as defined in the directive).

For example in Article 3 of the Medical Device Directive 93/42/EEC the essential requirements for "other" medical devices are defined as follows:

*Article 3*

*Essential requirements*

*The devices must meet the essential requirements set out in Annex I which apply to them, taking account of the intended purpose of the devices concerned.*

In case there is a potential risk, products which are "machinery" (according to Article 2a of the Directive 2006/42/EEC from May 17, 2006) must also comply with the essential health and safety requirements according to the Annex I of this directive in case these essential requirements are more specific than the essential requirements laid out in the Annex I of the Medical Device Directive 93/42/EEC.

At the beginning of Annex I of this directive, the general requirements for medical devices are stated as follows:

*ANNEX I*

*ESSENTIAL REQUIREMENTS I. General requirements 1. The devices must be designed and manufactured in such a way that, when used under the conditions and for the purposes intended, they will not compromise the clinical condition or the safety of patients, or the safety and health of users or, where applicable, other persons, provided that any risks which may be associated with their use constitute acceptable risks when weighed against the benefits to the patient and are compatible with a high level of protection of health and safety.*

(http://eurlex.europa.eu/LexUriServ/LexUriServ.do?uri=CONSLEG:1993L0042:20071 011: de:pdf)

---

**Exercise:**

Read Annex I of the Medical Device Directive 93/42/EEC, especially the details of the essential requirements for medical devices.

---

### 7.3. Structure of Essential Requirements

Annex I of the Medical Device Directive 93/42/EEC provides detailed information about the essential requirements. They are divided into 3 main groups:

- General requirements, which apply to all products and have to be met (93/42/EEC, Annex I: 1 – 6)
- Requirements regarding design and construction; requirements have to be met in case that they are applicable to the product (93/42/EEC, Annex I: 7 – 12)
- Requirements regarding information for all products that has to be provided by the manufacturer. Implementation of information or instruction can be handled product-specifically (e. g. product itself, label, packaging, instructions for use) (93/42/EEC, Annex I: 13)

The general requirements which are defined in 93/42/EEC, Annex I, 1 – 6 contain information on …

- The risks regarding safety
- How the product performance is achieved
- Conformity according the latest technical standards
- The product characteristics
- Performance (which has to be guaranteed over the defined product life as well as during storage and transport)

Moreover, a clinical assessment according to Annex 10 has to be conducted and documented.

The benefit-risk-analysis clearly has to prove that the benefit the product offers outweighs remaining risks.

The product-specific requirements regarding design and construction described in 93/42/EEC, Annex I, 7 – 12 are related to the following elements:

- Chemical, physical and biological requirements
- Information on infection or microbial contamination
- Construction and environmental properties
- Products with measurement function
- Protection from radiation
- Requirements for products that that are linked with an energy source or which contain one.

### 7.4. Harmonized Standards

Every directive has its essential requirements. Those are listed in Annex I. In order to comply with these requirements, manufacturers can make use of European "harmonized standards". Using these "harmonized standards" is voluntary. The manufacturer can refrain from these standards but in this case has to prove, that his product still complies with the essential requirements. Not using existing harmonized standards has to be justified by the manufacturer in a detailed and comprehensible way. He has to prove that he achieved the same level of safety as if he had used the harmonized standards.

## 7.5. Presumption of Conformity

The big advantage in using harmonized standards is the "presumption of conformity" which means if harmonized standards are used, it can be assumed, that a product meets the essential requirements. It is the manufacturer who assesses the conformity of his products regarding the essential requirements of the EU directive. The notified body then assesses the conformity of the quality management system of the manufacturer regarding the requirements of the EU directive. If harmonized standards for this product exist and when these are used, it can be assumed that the essential requirements are met (principle of assumption). Any presumption though is not a proof.

Harmonized standards referring to each of the European directives are published in the official gazette. An up-to-date list of the sources of every directive can be found on the following website: http://ec.europa.eu/enterprise/policies/european-standards/-harmonised-standards/index en.htm

## 7.6. Obligation of Documentation

It is the responsibility of the manufacturer to document that he has complied with the relevant essential requirements. This proof is part of the technical documentation and has to be provided as well as the whole technical documentation for potential assessment by the Competent Authority.

The technical documentation always has to be up-to-date concerning:
- Modifications of the product and/or of the manufacturing process
- Changes in the state-of-the-art technical development
- Standards and also new risks which have recently surfaced

The technical documentation has to be kept for at least 5 years, when implants are concerned even 15 years after placing the last product on the European market.

## 7.7. EU Declaration of Conformity

When the essential requirements are met, the manufacturer issues and signs a declaration of conformity. The manufacturer or his authorized European representative has to issue an EU Declaration of Conformity as defined in the CE directives of the "New Approach" (conformity assessment procedure). In the EU declaration of conformity all necessary references to the CE directives are included on which basis it was issued, as well as details about the manufacturer, his authorized representative and the product, the involvement (voluntary or required) of a Competent Authority, a list of the harmonized standards (voluntary) and further normative documents.

The CE directives of the "New Approach" put the responsibility for issuing an EU Declaration of Conformity in the hands of the manufacturer when the product is placed on the market. The EU Declaration of Conformity has to state that either the essential requirements of the CE directives are met or that the product complies with the defined construction details (type examination assessment document) and the essential requirements of the CE directives.

The EU declaration of conformity has to be kept for at least 10 years from the last date of production of the product – unless nothing else is stated in the CE directive. This again is the sole responsibility of the manufacturer. Sometimes, this responsibility is transferred to the importer or the responsible person who places the products on the market. The content of the EU Declaration of Conformity is defined in the CE directives regarding the specific products.

The declaration of conformity should consist of the following aspects:
- Name and address of the manufacturer
- Clear product description
- Source regarding the directive
- If applicable: name and address of the authorized EU representative
- Period of validity (starting point and endpoint)
- Date and signature

Also the following details may be included:
- Reference to the relating CE certificate
- List of implemented standards
- GMDN codes of the defined product
- Notified body that has issued the CE certificate

The "Blue Guide" of the European Commission ("Guide to the Implementation of Directives based on the New Approach and the Global Approach") lists details on the content of declarations of conformity (also compare the references in chapter 5.4. (http://ex.europa.eu/enterprise/policies/single-market-goods/files/blue-guide/guidepublic en.pdf)

## 7.8. Further Examples

There are several examples of checklists concerning the essential requirements of the directives on the worldwide web. On the websites of Notified Bodies one can find lists which might be very helpful in your day-to-day business routine.

## 7.9. Summary

The essential requirements define the mandatory basics for the protection of the public interest. Complying with the essential requirements is first priority. Only products that meet these requirements are allowed to be placed on the market. The content of the essential requirements depends on the potential risks of a product.

The essential requirements result in a high safety level. The requirements are derived from:

- Potential risks (physical and mechanical stability, flammability, chemical, electrical or biological properties, hygiene, radioactivity, precision) of a product
- The product and its performance (material, design, construction, manufacturing process, instructions supplied by the manufacturer)
- The defined protective objective (e. g. using a defining list)

It may also often be a combination of the above mention aspects. Therefore, several directives may be valid for a product because the essential requirements of more than one directive have to be met.

Applying the essential requirements depends on the potential risks of a product. Manufacturers have to conduct risk analysis in order to find out which essential requirements apply for the product. These analyses have to be documented and added to the technical documentation.

Essential requirements define the intended objectives and the risks that need to be prevented without mentioning the technical solutions that have to be applied to meet these requirements. This gives a lot of flexibility to the manufacturer to decide how to comply with these requirements. It also enables the manufacturer to adapt his choice of material and product design to the technological progress. Therefore directives of the New Approach do not need to be updated on a regular basis to the technological progress, because the assessment of the compliance of the requirements always takes place in the context of the latest technical developments at a given point of time.

The essential requirements are to be found in the annex of a directive. Although no detailed manufacturing processes are described in the essential requirements, the directives are more or less detailed in their descriptions. They should be as precise as needed to be legally binding (regarding to the national law) and enforceable. Moreover, the work of the European Standardization Organizations – employed by the Commission to define harmonized standards – has to be facilitated. The essential requirements have to be defined in such a way that the assessment of conformity to these requirements is still possible, even if no harmonized standards exist or in case that the manufacturer decides not to apply these harmonized standards.

## 7.10. Test Your Knowledge

| | |
|---|---|
| **Q1:** | Is it mandatory to apply harmonized standards in order to comply with essential requirements? |
| **A1:** | No, other standards can be applied as long as they guarantee the same level of safety as harmonized standards. |

| | |
|---|---|
| **Q2:** | Which information has to be contained in a Declaration of Conformity? |
| **A2:** | The Declaration of Conformity should consist of the following aspects: |
| | - Name and address of the manufacturer |
| | - Clear product description |
| | - Source regarding the directive |
| | - If applicable: name and address of the authorized EU representative |
| | - Period of validity (starting point and endpoint) |
| | - Date and signature |

> Also the following details may be included:
> - Reference to the relating CE certificate
> - List of implemented standards
> - GMDN codes of the defined product
> - Notified body that has issued the CE certificate

**Q3:** What is meant by "presumption of conformity"?

**A3:** This means that when harmonized standards are applied one can assume that the essential requirements of a directive (to which these standards refer) of the New Approach are met.

<u>7.11. References</u>

- Directive 93/42/EEC: http://eur-lex.europa.eu/LexUriServ/LexUriServ.do?uri=CONSLEG:1993L0042:2007101 1:de:PDF
- Liste of harmonized standards: http://ec.europa.eu/enterprise/policies/european-standards/harmonised-standards/index_en.htm
- 'Blue Guide' of the European Commission ("Guide to the implementation of directives based on the New Approach and the Global Approach"): http://ec.europa.eu/enterprise/policies/single-market-goods/files/blue-guide/guidepublic_en.pdf
- Website of a notified body with the possibility to download helpful documents: www.mdc-ce.de/1dloads.htm

## Chapter 8: Technical Documentation

*Dr. Stefan Menzl*

### 8.1. Learning Objective

This chapter describes the content and structure of a technical documentation for medical devices (depending on the classification).

### 8.2. Introduction

A technical documentation is mandatory – whether it is a product for clinical investigation, a custom-made product or a medical device of class I, IIa, IIb or III. The requirements are defined in the Medical Device Directive (MDD, Annex II 3.2(c) and 4.3, Annex III 3, Annex VII 3, and Annex VIII 3.1 and 3.2).

The objective of a standardized technical documentation by applying uniform methods is a clear description of straightforward and also complex facts. The Global Harmonization Task Force developed the so-called STED document (Summary Technical Documentation for Demonstrating Conformity to the Essential Principles of Safety). Moreover, a standardized documentation facilitates the necessary actualization and maintenance of the document.

The documents have to be straightforward in order to avoid misunderstandings and wrong interpretations. The structure must be clearly arranged but the structure and content can be arranged taking the target group into consideration.

### 8.3. Structure of a Technical Documentation

The Notified Body Group has developed a recommendation document that might be helpful (NB-MED/2.5.1/Rec 5, "Technical documentation", www.mdc-ce/downloads/R2_5_1-5_rev4.pdf).

The structure of a technical documentation or a design dossier consists of the following 12 chapters:

0    Application-form to the Notified Body

1    Introduction, description of the medical device

| | |
|---|---|
| 2 | Essential requirements checklist |
| 3 | Risk analysis |
| 4 | Drawings and design of the product, pictures and product specifications |
| 5 | Chemical, physical and biological testing |
| 5.1 | Bench-testing – pre-clinical testing |
| 5.2 | Biocompatibility testing |
| 5.3 | Biostability testing |
| 5.4 | Microbiological safety, material of animal origin |
| 5.5 | Coated medical devices |
| 6 | Clinical data |
| 7 | Qualification of packaging and shelf-life |
| 8 | Labelling |
| 8.1 | Labelling and instructions for use |
| 8.2 | Promotional material |
| 9 | Manufacturing process |
| 10 | Sterilization |
| 11 | Final assessment (risk-benefit assessment) |
| 12 | Declaration of Conformity of the manufacturer |

Moreover, it is worthwhile to add the following elements to the technical documentation:

- List of applied harmonized standards
  (http://ec.europa.eu/enterprise/policies/european-standards/harmonised-standards/ medical-devices)
- List of other applied standards
- Relevant literature and results of database research

## 8.4. Benefit of the Technical Documentation

The technical documentation is the proof for complying with the essential requirements according to the directive. Moreover, the correct definition of the medical device (IVD) and its classification is documented.

The technical documentation is a "living document" that has to be updated on a regular basis in order to reflect the latest technical developments as well as the design and manufacturing process of the product.

By using the technical documentation, the manufacturer can prove to the Notified Body and the authorities that its product meets the essential requirements.

## 8.5. Technical Documentation and Classification of Medical Devices

The examination of the technical documentation by the Notified Body varies, depending on the classification of a medical device and the details from the Annex of a directive.

If a medical device of class III is certified according to MDD, Annex II.3 and II.4, the Notified Body audits the whole quality management (QM) system of the manufacturer. It examines the documentation of the QM system, the QM manual and the corresponding procedures. The manufacturer has to present the design dossier to the Notified Body regarding the product-specific technical documentation. The Notified Body checks whether the essential requirements are met.

If a medical device of class IIa or IIb is certified according to MDD, Annex II, the Notified Body also audits the whole QM system of the manufacturer as well as the corresponding documents. The assessment of the technical documentation usually is part of the audit. The Notified Body also takes samples of the technical documentation of products in order to ensure that they comply with the essential requirements.

If a medical device of class IIb or III is certified according to MDD, Annex III, the Notified Body has to be provided with the technical documentation. The Notified Body also carries out a type examination. A certification is always issued with regard to the examination of the QM system according to Annexes IV, V and VI. In this procedure, the result of the type examination is the proof for complying with the essential requirements as it is described in the technical documentation. On top, the QM system guarantees a constant production according to the technical documentation.

Medical devices of class IIa can be certified according to MMD, Annex V in combination with Annex VII. The Notified Body conducts an audit of the QM documentation. A sample of the technical documentation will be examined as part of the audit.

If a medical device of class IIb or III is certified according to MDD, Annex V in combination with Annex III, the examination of the conformity with the essential requirements takes place parallel to each other. The audit of the QM system and documentation is carried out as well as the examination of the technical documentation.

Medical devices of class IIa can be certified according to Annex VII. This is possible due to an examination of the QM system according to Annex IV, V and VI. A sample of the technical documentation will be examined as part of the audit.

Also Annex VII of the MDD can be chosen to certify medical devices of class I. Only in this case – contrary to all other cases – the technical documentation will be reviewed by the Competent Authority (not by the Notified Body).

---

**Keep in mind:**

The technical documentation of class I medical devices that is certified according to Annex VII is conducted by the Competent Authority and not by the Notified Body! Monitoring by the authority and responsibility by the manufacturer guarantee in this case the compliance with the essential requirements.

---

The guideline of the European Commission provides worthwhile recommendations for the implementation of the directives of the New Approach and the Global Approach.

8.6. Technical Documentation File

It is the manufacturer´s obligation to establish a technical documentation that contains information on the design, manufacturing and the application of the product. The directives of the New Approach (also see chapter 7) also oblige the manufacturer to provide technical files as proof for complying with the essential requirements. These files can be part of the documentation of the QM system if the directive provides for a conformity assessment procedure on the basis of the QM system (Module D, E, H and its variants). The obligation starts on the day when the product is placed on the market.

In case the directive does not state otherwise, the technical files have to be stored for 5 years (starting the date of the last production) or 15 years in case of implantable devices. This lies in the responsibility of the manufacturer or its Authorized Representative (who resides in the EU). In some cases, the importer or the Authorized Representative for the import has to take over this responsibility.

The content of the technical files is defined in the respective directives. These files have to provide information on the design, manufacturing and the application of the device. The details included in the technical documentation depend on the nature of the product and on what is considered as necessary, from a technical point of view, for demonstrating the conformity of the product to the essential requirements of the medical devices directives. If harmonized standards have been applied, the technical documentation should also make clear where these have been used to demonstrate conformity with the particular essential requirements covered by the standards.

Some directives require that the technical documentation is written in an official language of a member state in which the procedures are planned to be conducted or in which a Notified Body is residing – or at least in a language accepted by the Notified Body. This should guarantee that the conformity assessment procedure is carried out in an orderly way by a neutral third party – in this case: the Notified Body.

A technical documentation is mandatory – whether it is a product for clinical investigation, a custom-made product or a medical device of class I, IIa, IIb or III. The requirements are defined in the Medical Device Directive (MDD, Annex II 3.2(c) and 4.3, Annex III 3, Annex VII 3, and Annex VIII 3.1 and 3.2).

As mentioned in the beginning of this chapter, the Notified Body Group has developed a recommendation document that provides good orientation on the general structure of a technical file and/or design dossier (NB-MED/2.5.1/Rec 5, "Technical documentation", www.mdc-ce/downloads/R2_5_1-5_rev4.pdf).

Experience with dealing with Notified Bodies shows that the recommended content of a technical documentation should **at least** contain:

- Table of contents
- Declaration of conformity of the manufacturer
- Description of the product and its planned variants
- Design specifications including the features of the raw material, performance and capacity limit of the product, manufacturing process as well as drawings and plans of construction components
- Assembly integrates and integrated circuits etc.
- Results of the risk analysis
- Results of calculations and examinations
- List of the applied (in full or on part) harmonized standards
- Proof that essential requirements are met
- Proof of compatibility with other products
- Clinical data
- Labelling of the instruction for use
- Relevant literature and results of database search

Supplier of raw material as well as sub-contractor can provide the Notified Body with documents to which the manufacturer of a medical device can refer. "White Label" manufacturers or "Original Equipment Manufacturers" can directly present documentation to the Notified Body.

The MDD enforces the responsibility of a manufacturer by asking a formal risk analysis for every product or every product group. In this case the harmonized standard EN ISO 14971 should be applied (see chapter 15 in the book).

---

**Exercise:**

Look up EN ISO 14971 and read the definition of this standard regarding the risk analysis.

---

According to the New Approach the European Commission asks the European Standardization Committee (CEN/CENELEC) to develop standards that prove the compliance of products with the essential requirements of a given directive. After examining if the essential requirements are successfully met by the standard, the European Commission published the title of the standard in the official EU gazette. By getting published, these standards become harmonized standards. If manufacturers apply these harmonized standards, it is presumed that the products comply with the essential requirements (MDD, Article 5). Nevertheless, the application of harmonized standards is voluntary. If manufacturers refrain from applying these harmonized standards, it is up to them to prove that the essential requirements are met.

Apart from the guidance document published by the Notified Body Group, some individual Notified Bodies offer internal directives about the content of the technical documentation, e. g. the following very detailed one:

0.  Name and address of the Legal-Manufacturer as well as his European representative

1.  Introduction, description of the product
- Short specification of the product
- Product history (e. g. modifications, further developments, previous model)
  - o  Intended use
  - o  Indication
  - o  Contraindication
  - o  Warning labels
- Accessories and packaging components
- Authorization of the distribution in other countries, e. g. FDA 510(k) or PMA clearance
- Planned modifications
- Classification of the product and its accessories under the Annex of the applied directive, f. e. MDD, Annex IX)
- Chosen conformity assessment procedure

2. Example

| Essential Requirements | Suitable? | Used standard | Conformity presented by | Reference to internal documents |
|---|---|---|---|---|
| 7.1 (text) | yes | ISO 10993 -1 -3 -5 etc | NAmSA test reports: - cytotoxity test (#xyz, dtd. 08/07/97 - 90 day implant (#xyz, dtd. 09/10/97 | Section 6.1 a) b) c) |

Tab. 8/1: Checklist Essential Requirements

3. Risk analysis according to EN ISO 14971

According to EN ISO 14971, a description in tabular form is accepted for: post-market surveillance data (history of adverse events), clinical experience and clinical risks. The same is also true for data collected under EN 12442 part 1-3 and MEDDEV 2.5-8 (risk management "tissue of animal origin")

The risk management file that describes the results of the risk analysis, should at least contain the following elements:

- General information
  - o Summary
  - o Reason of this document
  - o Content (description of all parts that were assessed), product description and intended use
  - o List of referenced documents (standards, specifications, design dossier, procedures)
  - o Definition of terms, abbreviations and acronyms

- Methodology
  - o In which project phase was the risk analysis conducted?
  - o What group of persons was part of the team that conducted the risk analysis as well as qualification and experience in the relevant medical area

- o Requirements regarding risk management activities
- o Hazard potential with normal use: risk analysis, potential hazard to patients/user (top-down-approach")
  - Identification methods of relevant risks (incl. sources of information)
  - Applied system to assess severity of the risk
  - Assessment method of potential causes of the risk
  - Applied system for assessing the probability of individual risks (frequency defined as "events per device")
  - Applied scheme combining severity and occurrence probability of risks and risk level
  - Acceptance criteria for every risk level

- o Hazard potential with malfunction
  - Identification methods of relevant risks (incl. sources of information)
  - Applied system to assess severity of the risk
  - Assessment method of potential causes of the risk
  - Applied system for assessing the probability of individual risks (frequency defined as "events per device")
  - Applied system for assessing the identification of every malfunction
  - Applied scheme combining severity and occurrence probability, identification and classification of risk level
  - Acceptance criteria for every risk level
- o Assessment procedure of information of the post-production phase
- o History of complaints and data from literature research
- Summarizing result report (date and signature)
  - o Hazard potential with normal use list of the potential risks, for every risk:
    - List of potential causes (weighted)
    - Risk assessment before risk minimization (severity, probability)

- Definition of the measures of risk minimization including definition of method (design, testing, production) as well as verification of the efficiency of the implementation
- Assessment of risk after risk minimization (severity, probability)
  - o Hazard potential with malfunction list of the potential risks, for every risk:
    - List of potential causes (weighted)
    - Risk assessment before risk minimization (severity, probability)
    - Definition of the measures of risk minimization including definition of method (design, testing, production) as well as verification of the efficiency of the implementation
    - Assessment of risk after risk minimization (severity, probability, identification and classification of risk level)
- List of the potential, reasonably foreseeable (max.) number of malfunctions as well as the causes (weighted)
  - Risk assessment before risk minimization (severity, probability, identification, classification of risk level)
  - Definition of the measures of risk minimization including definition of method (design, testing, production) as well as verification of the efficiency of the implementation
  - Assessment of risk after risk minimization (severity, probability, identification and classification of risk level)
    - o Assessment of new risks that could be the result of risk minimization measures
- Final assessment
  - o Completeness of the risk assessment
  - o Acceptance of the residual risks
  - o Risk-benefit assessment
  - o Signature of the responsible person and date

4. Drawings, design, product specification
   - Detailed description of the medical device

- Used materials and components
- Sample of the product
- Pictures
- Pre-production design control measures
- QMS (ISO EN 9001, ISO 13485) certificate of the design site
- Final product approval criteria including reference to verification and validation
- Sub-contractor for design and production
- List of the suppliers of critical components
- Treaties with sub-contractors
- Approvals of the authorities

5. Chemical, physical and biological testing

5.1. Bench-testing – pre-clinical studies
- Laboratory examinations (chemical and physical testing)
  - f. e. durability, pulling capacity, corrosion resistance, fatigue, long-term stability
  - Specific standards of product or product category has to be considered
- Laboratory examinations (chemical, biological, pharmacological, pharmacokinetic and toxicological tests
  - f. e. purity, toxicity, ADME (adsorption, distribution, metabolism)
- Performance tests
- Sterilization qualification
  - Does sterilization of the product influence its performance?
- Compatibility with medicinal products
  - Interactions between medicinal product and medical device
- Test protocols
  - Matrix of applied standards
    - Reason for not applied standards or elements of standards
    - Reference to test and validation
  - Tests of the final product

- o Tests on the accelerated aging or real-time on stability
- o Requirements regarding the tests for accelerated aging
- o For every test
    - ▪ Description of the test and the parameters
    - ▪ Test equipment
    - ▪ Test samples
    - ▪ Calibration
    - ▪ Acceptance criteria
    - ▪ Number of test samples and reason
- Result report of individual tests
    - o Deviations from protocol and reason
    - o Test data
    - o Statistical analysis
    - o Interpretation and conclusion
    - o Signature / date

5.2. Biocompatibility tests
- Applied standards
- Classification of the medical device in order to define applicable tests (see ISO 10993-1, table 1)
    - o Intended use
    - o Kind and duration of the contact with human body
- List of components and material that will get into contact with human body (directly or indirectly)
    - o Calculation of the total surface that get into contact with body or body liquids
    - o Description of the test samples (final product? Sterile? ...)
    - o Conducted tests
        - ▪ Reason for test selection
    - o Reason for applicable tests that were not conducted
- Conclusion
    - o Statement of the manufacturer that medical device is safe according to parameters tested
- Test results (result reports)

- Qualification of laboratories that conducted the tests
- Test samples
- Was the whole product tested or only its components?
- Specification of tested material
- Status of test samples
- Conducted tests (in detail)
- Applied standard
- Report of the test laboratory
- Test result (conclusive)

5.3. Biostability test

Influence of the biological matrix on the medical device or its components, e. g. corrosion, decline in coating

5.4. Microbiological stability

especially for medical devices which contain material/tissue of animal origin
Viral and bacterial safety, safety in regard to prions, reference to literature

EN 12442 part 1 - 3 (Animal tissue and their derivates used in the manufacture of medical devices,

- MEDDEV 2.11/1Application of Council Directive 93/42/EEC taking into account the Commission Directive 2003/32/EC for Medical Devices utilizing tissues or derivatives originating from animals for which a TSE risk is suspected (http://ec.europa.eu/health/medical-devices/files/meddev/2_-11_1_rev2_bsetse_january2008_en.pdf)
- MEDDEV 2.5-8 Guidelines on assessment of medical devices incorporating materials of animal origin with respect to viruses and transmissible agents www.meddev.info/_documents/2_5-8_____02-1999.pdf)
- Directive 2003/32/EC applies to materials from bovine, ovine, caprine species, deer, elk, cat, and mink only, http://eur-lex.europa.eu/-LexUriServ/LexUriServ.do?uri=OJ:L:2003:105:0018:0023:EN:PDF)

The manufacturer must consider the following aspects and provide relevant documents if needed:

- o Reason/justification of using material of animal origin or tissue of animal origin or derivates
- o Original materials
- o Species from which material originated
- o Assessment of clinical benefit versus residual risks (possible choice of alternative material)
- o Studies for elimination or inactivation of BSE/TSE-relevant substances
- o Applied risk analysis
- o Applied measures for infection risk minimization
- o Control of original material, final product and supplier
- o Planned audit measures (suppliers of raw material)
- o EDQM-certificate, if available

## 5.5. Coated medical devices

For example:

- Heparine-Coating
- Silver- / Gold-Coating
- Pyrolytic Carbon Coating
- MPC-ML-Coating (Methacryloyl Phosphoryl Choline Lauryl Methacrylate)
- Parylene Polymer Coating
- Collagene / Gelatine Coating
- PEG Coating (Polyethyleneglycol as Lubrication)
- E-Beam Treatment (Cross linkage)
- Titanium / HA Spray-Coating

Requirements to performance and product safety (stability of coating in the biological matrix)

- Hydrophilicity
- Microbiological assessment
- Fibrinogen adsorption
- Adhesion of blood platelets / activation cascade
- contact activation tests

6. Clinical assessment

The following sources can be considered suitable and mandatory for the clinical assessment:

- MDD (93/42/EEC) - Annex 10
- ISO EN 14155-1 and -2
- MEDDEV 2.7.1
- GCP (good clinical practice) requirements

European Legislation

- Council Directive 90/385/EEC of 20 June 1990 concerning active implantable medical devices
- Council Directive 93/42/EEC of 14 June 1993 concerning medical devices

GHTF final documents

- SG1/N011:2008 Summary Technical Documentation for Demonstrating Conformity to the Essential Principles of Safety and Performance of Medical Devices (STED)
- SG1-N44:2008 Role of Standards in the Assessment of Medical Devices
- SG1/N029:2005 Information Document Concerning the Definition of the Term "Medical Device"
- SG1/N040:2006 Principles of Conformity Assessment for Medical Devices
- SG1-N41R9:2005 Essential Principles of Safety and Performance of Medical Devices
- SG5/N1R8:2007 Clinical Evidence – Key definitions and Concepts
- SG5/N2R8:2007 Clinical Evaluation

International standards

- ISO 14155-1 Clinical investigation of medical devices for human subjects – Part 1: General requirements
- ISO 14155-2 Clinical investigation of medical devices for human subjects – Part 2: Clinical investigation plan
- ISO14971 Medical devices – application of risk management to medical devices.

European guidance documents

- MEDDEV 2.10/2 Designation and monitoring of Notified Bodies within the framework of EC Directives on medical devices
- MEDDEV 2.12/2 Guidelines on post-market clinical follow up
- NBOG BPG 2009-1 Guidance on design dossier examination and report content www.nbog.eu/resources/NBOG_BPG_2009_1.pdf
- NBOG BPG 2009-4 Guidance on NB's Tasks of Technical Documentation Assessment on a Representative Basis www.nbog.eu/-resources/NBOG_BPG_2009_4_EN.pdf

A clinical assessment is mandatory for all medical devices of all classes (MDD). Clinical studies have to be conducted for implantable medical devices and medical devices of class III (MDD, Annex X 1.1a). Clinical investigations have to be conducted also, for devices of all other classes in case available clinical data does not suffice.

Clinical studies are necessary if MEDDEV 2.7.1 applies:

- Completely new medical devices
- Unknown components or performance characteristics
- Unknown mode of action
- Significant modification of an existing medical device (influence on safety and/or performance)
- New indication of an existing medical device
- Use of new materials that come into contact with the human body
- In case an existing medical device is used significantly longer than defined earlier

Together with the final report of the clinical study, the following parts should also be added to the technical documentation: the study protocol, the statement of the ethics commission and the approval of the Competent Authority.

It is mandatory to conduct a critical assessment of all results and findings of the study. In case the study is not finalized, foreseeable results and the planned study end date should be mentioned.

The following documents should be available for the clinical assessment:

- Final report according to MEDDEV 2.7.1
- Copies of all literature which are referenced in the clinical assessment
- Search strategy for literature as well as inclusion/exclusion criteria
- Analysis of clinical risks (maybe as part of the overall risk analysis)
- Post-market surveillance (PMS) data if available (or PMS data from previous model)
- Study protocol, final report as well as report of preclinical studies
- Clinically relevant results of non clinical investigations
- CV and signature of a competent clinical expert

7. Qualification of packaging and durability

In this part, the following tests have to be conducted:

- Physical examination of the packaging
- Performance test of a medical device regarding real-time aging as well as accelerated aging
- Proof of preservation of sterilization over the product life

The following documents can be applied

- EN 868 -1 ff Packaging for terminally sterilized medical devices
- ISO 11607 Packaging of medical devices that are sterilized with the package
- ASTM D999 -tests
- NAmSA Dust Drum Tests
- Real time aging
- Q10 - accelerated aging test

For sterile medical devices the following records should be added to the technical documentation:

- Description of the packaging and of the specification of the packaging material

- Certificate of the supplier
- Proof that packaging material can be used for the applied sterilization method
- Biocompatibility of the packaging material if needed
- Packaging integrity test
- Examination of the microbial barrier
- Labelling
- Seal strength test
- Durability test / aging
- Shipment simulation test, vibration, drop and roll test
- Process validation report of the packaging process

8. Product labelling, package insert, information for patients and users, promotional material

- Conformity with the requirements of the product labelling and instructions for use according to Directive 93/42/EEC, Annex I, 13. Supply of information by the manufacturer as well as Directive 90/385/EEC, Annex I, 11 ff.
- Conformity with EN 980, EN 1041 and ISO 15223
- All elements of the product labelling and instructions for use

---

**Exercise**

Promotion is key to the distribution of medical devices – but at the same time a highly sensitive issue. Look up the topic "promotional material" in the above mentioned documents.

---

9. Production

Description of the manufacturing process (incl. flow diagram)

- Proof of complying with harmonized standards, f. e. FS 209E, ISO 14644, ISO 14698
- Certificate of the quality system (EN ISO 9001, ISO 13485)
- EC certificate according to MDD, Annex II.3 (full quality assurance system)

- Control of the labelling
- Traceability of materials and products
- Bioburden of the product as well as the environment (particle)
- Pyrogene test
- Preventive monitoring of manufacturing processes
- Viral and Prions activation processes

10. Sterilization

EN 550 series and ISO 11130 series describe in detail:

- Description of the qualification of the sterilization process as well as a summary of the validation (the chosen method must have a sterility assurance level of at least 10-6)
- Process validation (report) including proof of physical and microbial performance
- Proof of certification of the sterilization site by a Notified Body (ISO 9001/2, ISO 13485, 13488, EN 550 series, 11130 series)

11. Conclusion

This is the part where a risk-benefit statement has to be given.

12. Declaration of conformity (draft version if final version is not yet available)

Example of a declaration of conformity

Annex

Example: European standards as well as other documents referring to technical documentation

| Document Number | Title of Document |
|---|---|
| EN ISO 9001 | Quality Systems |
| ISO 13485 | Particular requirements for the application of ISO 9001 |
| EN 550 | EtO Sterilization |
| EN 552 | Irradiation Sterilization |
| EN 554 | Sterilization by moist heat |

| EN 556 | Sterilization of medical devices. Requirements for medical devices to be designated "STERILE". |
|---|---|
| ISO 14155 | Clinical Investigations of medical devices |
| ISO 11134 | Sterilization of health care products – Steam Sterilization |
| ISO 11135 | Sterilization of health care products – EtO Sterilization |
| ISO 11137 | Sterilization of health care products – radiation sterilization |
| ISO 10993 part 1 | Biological testing of medical devices – general requirements |
| ISO 10993 part 5 | In-vitro tests for cytotoxicity |
| ISO 10993 part 11 | Tests for systemic toxicity |
| EN 980 | Terminology, symbols for use in Medical Device labels |
| ISO 15223 | Symbols to be used in Medical Device labels, labelling and information to be supplied |
| EN 1041 | Terminology, symbols and information provided with medical devices – information supplied by the manufacturer with Medical Devices |
| ISO 14971 | Application of risk management to medical devices |
| EN 868 part 1 | Packaging materials and systems for medical devices which are to be sterilized – Part 1: General requirements and test methods |
| ISO 14644 | Cleanrooms and associated controlled environments |
| ISO 14698 | Cleanrooms and associated controlled environments – Biocontamination |
| USP | United States Pharmacopeia |
| E Ph | Pharmacopeia Europaea |
| EN 45014 | General criteria for suppliers declaration of conformity |
| MEDDEV 2.12/1 | Guidelines on a Medical Devices Vigilance System |
| NB-MED/2.5.2/Rec2 | Reporting of design changes and changes of the quality system |
| MEDDEV 2.7.1 | Evaluation of Clinical Data |

Tab. 8/2: European Norms and Standards as well as other documents referencing to Technical Documentations

Further information is to be found in ISO 16142 Medical devices – Guidance on the selection of standards in support of recognized essential principles of safety and performance of medical devices.

> **Exercise:**
>
> Go to ISO 16142 Medical devices – Guidance on the selection of standards in support of recognized essential principles of safety and performance of medical devices and check the above mentioned statements.

## 8.7. Summary

The essential requirements define the necessary elements for the protection of the public interest. The compliance with these requirements is mandatory and evidence is delivered in the technical documentation. The essential requirements provide content and structure for the creation of the technical documentation.

Several guiding documents are available regarding this structure and the required content of the technical documentation.

## 8.8. Test Your Knowledge

> **Q1:** What is the benefit of a technical documentation?
>
> **A1:** The technical documentation is the proof that the essential requirements (according to a defined directive) are met. Moreover, in the technical documentation the decision path of the classification of the medical device (IVDs, etc.) can be easily followed, thus making sure that the product was correctly classified.

> **Q2:** A medical device has been designed and has not been changed in any way for 4 years. Does the manufacturer during this time have to work on the technical documentation?
>
> **A2:** Yes, he has, as the technical documentation is a "living document". Both – the risk assessment as well as the clinical assessment – are parts of the technical documentation that have to be updated on a regular basis. Therefore, the technical documentation has to be updated regularly as well.

| **Q3:** Does the Notified Body examine the technical documentation of class I products (not sterile, no measurement function) during the audit or does the documentation has to be sent to the Notified Body beforehand? |
|---|
| **A3:** Neither nor. The examination of the technical documentation of class I products (not sterile, no measurement function) is carried out by the local authority, not by a Notified Body. |

## 8.9. References

- www.mdc-ce.de/downloads/R2_5_1-5_rev4.pdf
- http://ec.europa.eu/enterprise/policies/european-standards/harmonised-standards/medical-devices
- http://ec.europa.eu/health/medical-devices/files/meddev/2_11_1_rev2_bsetse_january2008_en.pdf
- www.meddev.info/_documents/2_5-8____02-1999.pdf

## Chapter 9:    Norms and Standards

*Dr. Sibylle Scholtz*

### 9.1. Learning Objective

This chapter focuses on the fact that the application and compliance with national and international standards does make sense and is in fact necessary when certifying medical devices and in vitro diagnostic (IVDs).

### 9.2. Introduction

The word "norm" is defined in Pfeiffer (Pfeiffer W, Etymologisches Wörterbuch des Deutschen, dtv, 1997) as "a rule, guideline, directive for size (for technical devices), a guideline (for workload, use of material) defined work performance, legislation, model ... ." The word is derived from the Latin word "norma" that can be translated "angular, guideline, rule, directive." Related terms are "to nominate, to norm, standardization or normal and to normalize."

Standards rule our lives – as far as social interaction is concerned but also regarding to technology, environment and health.

Standards do play an important role for medical devices. They describe the latest technical developments and are usually far more often revised compared to legal directives. By complying with European harmonized standards a "presumption of conformity" is achieved. This makes it easier for the manufacturer to prove compliance with the essential requirements according to MDD. This also helps the manufacturer placing innovative products much faster on the market. As far as medical devices and IVDs are concerned, norms define processes. They define a recognized standard although they are not legally binding directives. Several countries have institution which take care of this topic: in Germany it is the DIN (Deutsches Institut für Normung), in Europe EN (European Norms) are developed by CEN, CENELEC or ETSI and globally the ISO (International Organization for Standardization).

International
Organization for
Standardization

Fig. 9/1 and 9/2: Logos of DIN (www.din.de) and ISO (www.iso.org)

9.3. European Norms

The development of uniform European standards is due to the necessity to prevent barriers to trade that arose by the application to different national standards. It is also necessary to create a framework that reflects the increasing international trade relations. Of course, it is a challenge to transfer these international standards into national directives. The New Approach was helpful when implementing international directives (see text 1985 published in the official gazette of the EU: ABl. EEC, 1985, No C 136/01). The objective of this New Approach was not only to foster the free trade of merchandise within the internal European market but also to achieve a high safety level for patients, users and third parties, the minimization of product risks and legal security within Europe. On the basis of the New Approach, technical product specifications – that are mentioned in the European directives for medical devices – are described in detail in the harmonized standards. Nevertheless, the application of harmonized standards is voluntary. Applying these harmonized standards results in the presumption of conformity (= complying with the essential requirements of the MDD). Because of the globalization process, European standards (DIN EN) are more and more replaced by international standards (DIN EN ISO) that facilitates the global market access for manufacturers. (source: BVMed, Berlin, www.bvmed.de).

The Lisbon Treaty is a treaty binding under international law between the 27 member states of the European Union that was signed on December 13, 2007 and came into effect on December 1, 2009. Article 114 resp. 115 AEUV (setting out rules governing cooperation within the EU) is the basis according to which legislative and administrative directives of the member states are harmonized in order to avoid distortions within the European internal market. Up to now there are more than 20 European directives which were published and are mandatory, for example the directive for medical devices 93/42/EEC and the IVD directive 98/79/EEC. These directives have to be transferred by the EU member states as well as the EFTA countries into national law. Product specific details are defined via European standards (EN). Their development originates from a mandate by the EU. In this area, a mandatory transfer into national standards of the EU and EFTA countries is planned.

As EU directives are restricted to the essential safety requirements of a defined product group, these directives are described in detail by harmonized standards. Applying these harmonized standards leads to a presumption of conformity, nevertheless applying these harmonized standards still is voluntary. The European organizations like CEN (Comité Européen de Normalisation), CENELEC (Comité Européen de Normalisation Électro-technique) and ETSI (European Telecommunications Standards Institute) were asked to develop European standards.

Fig. 9/3, 9/4 and 9/5: Logos of CEN, CENELEC and ETSI (www.cen.eu, www.cenelec.eu and www.etsi.org)

The national participants of CEN and CENELEC (for example in Germany: DIN) are obliged to implement the European standard without changes. In case there are national standards which are inconsistent with these European standards, these national standards are to be withdrawn. If European harmonized standards are correctly applied by the manufacturer, the Notified Body will – according to the presumption of conformity – assume that the essential requirements are met by the product and therefore the product is allowed to be placed on the market. This is true as long as this EN is listed as a harmonized standard in the official EU gazette and when at least one member state has already transferred this norm into national standard. If a manufacturer refrains from applying these harmonized standards, he needs to refer to another directive. Then it is up to the manufacturer to prove (burden of proof!) that his products meet the essential safety requirements.

Regarding the directive 98/34/EEC a harmonized standard is a technical specification which was developed by CEN or CENELEC. Standards are called "harmonized" when they are published in the official EU gazette, when the normalization mandate is mentioned in the preface of the text or when the reference to the norm is described in detail in the annex ZA.

---

**Exercise:**

Look up information on the "New Approach" in the worldwide web.

---

9.4. International Standards

International standards reduce barriers to trade. It has to be mentioned though that ISO standards have not the same meaning as EN standards within the EU. When applying ISO standards, there is no presumption of conformity. The ISO (International **O**rganization for **S**tandardization) is an international association of organizations dedicated to normalization. They develop international standards in nearly all areas except for electrical engineering (responsible: International Electrotechnical Commission, IEC) and for telecommunication (responsible: International Telecommunication Union, ITU). The World Standards Organi-zation (WSC) consists of these three organizations: ISO, IEC and ITU.

Fig. 9/6, 9/7 and 9/8: Logos of ISO, IEC and ITU, which form WSC (www.iso.org, www.iec.ch and www.itu.int)

There are efforts to connect ISO and EN standards. These harmonized standards are recognized by the reference "EN ISO". If such an EN ISO standard exists and its corresponding EN standard is harmonized, it can be helpful to refer to it because this standard is accepted in other parts of the world (outside of the EU).

Two agreements are the basis of these EN ISO standards: The Vienna Agreement (1991) that defines the cooperation of CEN and ISO and the Dresden Agreement

(1996) which created the necessary framework of cooperation between IEC and CENELEC.

In this process both can happen: international standards can become European standards and vice versa.

## 9.5. Application of Standards – Assessment of Conformity of Medical Devices

The first step of the assessment of conformity of medical devices is answering the question whether or not a directive applies (for example medical devices or IVD or other products like medicinal products, cosmetics or foods).

If a medical device represents a potential risk to the human body has to be decided as well. There are several risk classes (e. g. see § 13 MPG – German Medical Devices Act). The definition of the intended use is solely up to the manufacturer, as well as the choice of the conformity assessment procedure.

At the beginning of this process, inquiries should always be conducted for existing harmonized standards. On the same level as harmonized standards there are also the monographs of the Pharmacopoea Europaea.

The next step consists of a checklist of the essential requirements that need to be met. This checklist is a part of the technical documentation as well as for the declaration of conformity. It also represents the basis for the CE certification after assessment by the notified body.

---

**Exercise**

Have a look at the following websites regarding further information on harmonized standards:

- European Commission: http://ec.europa.eo/enterprise/policies/european-standards/documents/harmonised-standards-legislation/list-references
- Official gazette of the EEC: http://eur-lex.europa.eu
- DIN: www.din.de
- BfArM: www.bfarm.de
- DIMDI: www.dimdi.de/static/de/mpg/index.htm

---

- Engineering association of technical communication: www.ce-richtilinien.eu/richtlinien/MedProd.html
- CEN, CENELEC, ETSI, EC and EFTA: www.newapproach.org
- International Medical Device Regulators Forum (IMDRF): www.imdrf.org
- UN Economic Commission for Europe: www.unece.org

## 9.6. Revision of Harmonized Standards

When harmonized standards are revised, the final versions are published in the official gazette of the EU with all the necessary details (reference to the revised standard, the outdated version as well as the date when the new version comes into effect. Revisions take place because of the latest technical developments or because – when applying the harmonized standards – parts of it turn out not to be feasible. Also the new edition of general standards are that kind of revisions, for example the third edition of EN 60601-1 (Medical electrical devices, part 1: general safety definition).

The revision of the directive 93/42/EEC which resulted in the EU directive 2007/47/EEC which came into effect on March 21, 2010 led to the revision of all harmonized standards related to it.

As far as EN standards were concerned, annex ZA was revised (informative part which explains the relation between standard and legal requirement). Moreover, the text of the standard was revised.

Regarding the EN ISO standards, only annex ZA was revised. A so-called gap analysis was carried out to identify essential requirements that were not or only to a certain extent met. These gaps were then closed in the meetings of the ISO and IEC.

## 9.7. Summary

Harmonized standards explain essential requirements of a directive in detail. By applying harmonized standards, a presumption of conformity can be achieved. The New Approach facilitates the legislation and makes it easier to keep up with the ongoing technical developments in this field. Although the New Approach does not define detailed specifications, it guarantees a high level of safety for patients, users and third parties. The definition of details via standards enables a fast modification to the latest technical developments – the definition of the standards is up to the experts.

This legislation gives the manufacturer a certain amount of freedom resulting in an easier international market access.

On the other hand, a lot of responsibility lies in the hand of the manufacturer regarding the safety of the product over the whole product life. Only if a certain level of potential danger of the product for human beings is reached, an authorized inspection office has to be called in. Further official controls take place when products are already placed on the market.

It has to be noted that these compilations of norms have a practical side, but are often somewhat "clumsily" and do not always reflect the reality.

The standardization of norms and certification procedures leads to a reduction in costs because double assessments are avoided. A global harmonization is worthwhile especially for international companies. Nevertheless, there will be controversies and disagreements in the future about what is looked upon as appropriate or state of the art.

## 9.8. Test Your Knowledge

| | |
|---|---|
| **Q1:** | Which association works on a global level on (inter)national directives? |
| **A1:** | World Standards Cooperation (WSC) |

| | |
|---|---|
| **Q2:** | Why should harmonized standards be applied? |
| **A2:** | Applying harmonized standards will result in a presumption of conformity. |

| | |
|---|---|
| **Q3:** | Is it mandatory for a manufacturer to apply harmonized standards? |
| **A3:** | No. |

## 9.9. References

- European Commission: http://ec.europa.eu/enterprise/policies/european-standards/documents/harmonised-standards-legislation/list-references
- Official EU gazette: http://eur-lex.europa.eu
- DIN: www.din.de
- ISO: www.iso.org

- CEN: www.cen.eu
- CENELC: www.cenelec.eu
- ETSI: www.etsi.org
- IEC: www.iec.ch
- ITU: www.itu.int
- BfArM: www.bfarm.de
- DIMDI: www.dimdi.de/static/de/mpg/index.htm
- Engineering association of technical communication: www.ce-richtlinien.eu/richtlinien/MedProd.html
- www.newapproach.org
- International Medical Device Regulators Forum (IMDRF): www.imdrf.org
- UN Economic Commission for Europe: www.unece.org
- www.din.de/sixcms_upload/media/2896/D_Freizeit2012.pdf
- www.din.de/sixcms_upload/media/2896/D_Essenwimmel2012.127729.pdf
- www.iso.org/iso/wsdposter2005.pdf

## Chapter 10: Regulatory Intelligence
*Dr. Carsten Rupprath*

### 10.1. Learning Objective
In this chapter you will learn about the important tasks of the function "regulatory intelligence" (RI) – especially within global players. It will also be described how to cope with organizations, changing standards and norms in an increasingly complex regulatory landscape.

### 10.2. Introduction
"Nothing is as constant as change" Heraclitus of Ephesus once said. Back then he did not think about the registration of medical devices, but his saying hits the nail on the head. International requirements for the registration of medical devices are constantly changing. It can be devastating for a MedTech company not to be aware of these changes and risk the sales of their products. Therefore Regulatory Intelligence is a key business success differentiator for MD companies and pharmaceutical companies. Regulatory requirements set by authorities have to be met; state-of-the-art technology must be anticipated to guarantee a broad and sustained market access. Quite often – currently for example in the Middle East – new regulatory systems are being implemented, because these countries strive to regulate and control the health care market in order to protect their people. They usually copy parts of the existing regulatory systems of the leading Western industrial nations and adapt these to their national circumstances.

### 10.3. How to Keep up with Worldwide Regulatory Changes?
Regulatory requirements are often modified and constantly changing. These changes have to be recognized by the companies as soon as possible in order to comply in time. Therefore it is important for a company to have Regulatory Intelligence employees who closely observe introductions, revisions or changes in the global Regulatory landscape that can affect regulatory strategy in the different countries. If these changes are not recognized and taken care of in time, the market access of the products can be in danger or even non-compliant products have to be taken off the market.

There is a variety of information channels for a company to stay abreast of changes in the regulatory landscape, e. g. new or revised regulations:

- An important source on new regulations are the websites of the authorities, e. g. in the EU it is the website for medical devices of the European Commission (http://ec.europa.eu/health/medical-devices/documents/index en.htm)
- Another source are trade associations consisting of companies who jointly address their interests with policy makers. The following associations should be mentioned:
    - In Germany: BVMed (www.bvmed.de) and SPECTARIS (www.spectaris.de)
    - On a European level: Eucomed (www.eucomed.org) and especially for contact lenses and contact lenses care products EUROMCONTACT (www.euromcontact.org)
    - Comparable industrial associations (of MD manufacturers) exist worldwide which regularly inform their members on new regulations, e. g. ADVAMED in the US (www.advamed.org) or MECOMED (www.mecomed.org) in the Middle East.
    - There are also professional associations like the Regulatory Affairs Professional Society (www.RAPS.org) or the Organization for Professional in Regulatory Affairs (www.topra.org) where Regulatory Affairs employees are organized. These organizations gain and analyze publicly available regulatory information and communicate these to their members.
- A valid source of information is also a Notified Body. When meeting with a Notified Body to discuss the regulatory strategy concerning the registration of a product, this meeting can also be used for an exchange on new regulations.
- Gathering information which is displayed on the website of the local authority on a regular basis is also a good source of information, e. g. The BfArM website in Germany (www.bfarm.de), AGES website in Austria (www.ages.at) or the healthcare authority website of the Arab Emirates (www.moh.gov.ae/en).
- In global companies, information exchange about new regulations between departments is important as employees at e.g. Quality or R&D might have different information and it is possible to discuss the implications of changed regulations or standards from different angles.
- Information are often also provided by Regulatory Affairs magazines or Medical Device journals, e.g. the "Journal of Medical Device Regulation",

"Medizinprodukte-Journal", "Medizinprodukterecht", "Clinica" or "The Gray Sheet".

- External agencies offer trainings on Regulatory Affairs topics or market access.
- Taking part in the work of organizations defining, developing or revising norms, gives firsthand insight of this process.
- There are also a large amount of free-lance Regulatory Affairs consultants who support companies in their market access activities. The information provided by these consultants should be carefully looked at, because freelancers have sometimes a tendency to exaggerate regulations for obtaining bigger orders from their customers.
- Even recalls of a product or the so-called FDA Warning Letters or internal sources of a company can give insight to new regulations. The "FDA Warning Letters" are publicly available in the internet (www.fda.gov/ICECI/EnforcementActions/WarningLetters/ default.htm)

It is important to know where to look (use several sources of information!) and how to use the information that has been gathered. As Regulatory Intelligence is a key business success differentiator, an operating procedure that defines the structure of this process is of great importance for a company.

---

**Exercise:**

Look up the website of the national health authorities in the Middle East and Oceania (A little hint from the author: the Saudi authority is called Saudi FDA and the Australian authority is abbreviated TGA).

---

10.4. Summary

"Panta rhei" (everything is moving) – this saying is also attributed to Heraclitus of Ephesus. This is also true for the current regulatory environment.

Regulatory Intelligence employees are on the one hand responsible for setting up a storage system (e.g. database) where all the local requirements for the registration of medical devices can be found. On the other hand they have to check on a regular basis, if these regulations are still valid. They have to search and collect new information, review, summarize and analyze it ... and finally interpret the gathered information and integrate the details in the storage system.

By closely observing the national legislation regarding new regulations for medical devices or medicinal products, Regulatory Intelligence takes care of a sustained market access. As the general trend in the EU but also worldwide is towards diversification of regional regulations, it has to be stated that Regulatory Intelligence is a key business success differentiator.

## 10.5. Test Your Knowledge

| | |
|---|---|
| **Q:** | What consequences can arise for a MedTech company if Regulatory Intelligence is neglected? |
| **A:** | The market access of its medical devices could be affected or even prevented. Regulatory Intelligence helps to react to new regulatory requirements and to produce products that comply with the regulations and therefore can be distributed in a market. |

## 10.6. References

- The Medical Device Directive 93/42/EEC dated June 13, 1994
- www.bfarm.de
- www.ages.at
- www.moh.gov.ae/en
- http://ec.europa.eu/health/medical-devices/documents/index_en.htm
- www.bvmed.de
- www.spectaris.de
- www.eucomed.org
- www.euromcontact.org
- www.mecomed.org
- www.advamed.org
- www.RAPS.org
- www.topra.org
- www.fda.gov/ICECI/EnforcementActions/WarningLetters/default.htm

## Chapter 11: Criteria for the Selection of a Notified Body (NB)

*Dr. Stefan Menzl, Myriam Becker*

### 11.1. Learning Objective

Manufacturers of medical devices have to take into account several criteria when choosing a Notified Body. Selecting the Notified Body, which fits best in regard to the medical devices of a company, significantly increases the chances for a successful registration. This chapter focuses on the criteria for the selection of a Notified Body and how to find the appropriate Notified Body.

### 11.2. Introduction

Every manufacturer of medical devices who is planning to promote his products within the European Union (EU) has to use the services of a Notified Body to get a CE-mark. In order to get a CE-mark, the medical device has to comply with the requirements of the relevant EU directive and in particular with the essential requirements on safety and performance.

The methods which a manufacturer may use to demonstrate that a medical device complies with applicable requirements varies – depending

- On which EU directive applies (medical device directive 93/42/EEC dated June 14, 1993; the active implantable medical device directive 90/385/EEC dated June 20, 1990; the in vitro diagnostic medical device directive 98/79/EEC dated October 27, 1998) and
- On the device classification.

The annexes of the directives define a variety of conformity assessment procedures. The Notified Body must be qualified to perform all functions defined in the annexes for which it is designated.

### 11.3. Definition of a NB

A Notified Body is an organization, accredited by a national authority of an EU member state and recognized by the European Union to audit quality systems, test medical devices, and assess technical documentation. This is to confirm that the manufacturers and their medical devices meet the applicable directives (e.g. MDD) or standards (e.g. ISO 13485). Upon successful completion of an audit, medical device testing, or assessment of a design dossier, the Notified Body issues a certificate or assessment report that confirms compliance with the applicable European laws.

The following activities are usually performed by a Notified Body:

- *Full Quality Assurance:*

  The Notified Body performs an assessment of the manufacturer's quality system. Sampling across a range of products and processes will be looked at to ensure that requirements are met.

- *Design examination:*

  The Notified Body assesses the design dossier for an individual product to assure compliance with the essential requirements.

- *Type examination:*

  The Notified Body assesses the technical documentation (technical file) and performs product tests for every type of product.

- *Production and Product Quality Assurance:*

  The Notified Body performs an assessment of the manufacturer's quality system covering the production or inspection (Production QA) or final inspection (Product QA). Sampling will occur across the range of products to ensure that relevant technical files are available and relevant processes are undertaken to meet the requirements.

Currently, there are 75 Notified Bodies accredited for Directive 93/42/EEC (MDD) (http://ec.europa.eu/enterprise/newapproach/nando/index.cmf?fuseaction=directive.n otifiedbody&dir id=13).

Manufacturers of medical devices are free to apply to any Notified Body in the EU that is accredited to carry out the desired conformity assessment procedure for the selected product regardless of the member state in which the Notified Body is established. This means that member states have to accept products assessed by foreign Notified Bodies to their national market. Also Notified Bodies have to accept CE assessments from other Notified Bodies, which is important as this allows trans-ferring CE certificates from one Notified Body to another.

From the standpoint of companies Notified Bodies are both – consultants and contractors. Accordingly, they operate under a formal contract and are paid for their service. It would be rather shortsighted to select a Notified Body only because of the price. A manufacturer of medical devices should wisely choose his Notified Body because he will probably have to closely work together with this Notified Body for many years to come.

11.4. Accreditation and Experience of a Notified Body

The most important criteria regarding the selection of a Notified Body is its ability to certify all types and classes of medical devices that a company plans to promote in the European market. This is especially important if the company offers medical devices ranging from class I to class III, and even more so if the company also pro- duces combination products, for example devices combined with tissue, blood- derived products or medicinal products.

A drug-device combination product requires a "consultation procedure" with one of the European drug agencies. In this case it is helpful when the Notified Body of choice is experienced with this kind of procedure.

In case that medical device consists of material of animal origin and therefore might be classified as a class III product, the Notified Body should have experience with this kind of assessment. For these products the submission and assessment of the design dossier and consideration of directive 2003/32/EC (utilizing tissues of animal origin) is inevitable. Before granting a CE-mark, the Notified Body has to draw up a summary evaluation report and submit this report to the national regulator of materials of animal origin. The national regulator then calls for the opinion of the other member states regarding the assessment report of the Notified Body.

---

**Exercise:**

See chapter 5 "Special Regulations for Medical Devices" for further information on the distinction between medicinal products and medical devices, consultation procedures, the usage of material of animal origin and their consequences regarding the CE registration.

---

Moreover, it can be of advantage for a medical device manufacturer, when the Notified Body shares its experience through best practices in the market or offers in-depth trainings. Such kind of firsthand information help the manufacturer to better understand how the Notified Body interprets laws, regulations, standards and general requirements.

It should be considered that not all Notified Bodies have the same level of influence within the Notified Body Operations Group (NBOG, www.nbog.eu) or the same reputation with the national authorities.

Manufacturers should be well aware of these differences and select a Notified Body that is well recognized and able to successfully defend its opinion against other Notified Bodies or relevant authorities in case that the company is challenged.

### 11.5. NB: Size and Staffing

Generally speaking a Notified Body with a very broad range of accreditation will be able to meet many needs a manufacturer is likely to face regarding future product development activities. All of the "bigger" Notified Bodies have experience with certifying products of various types, classes or combination products.

Smaller Notified Bodies may have their limitations if confronted with highly speciali-zed products. Also, they might not necessarily be the best choice if a company's future plans involve expansion to Japan or Canada because in these countries special accreditation of the Notified Body is needed. Not all Notified Bodies have a worldwide presence, and some might not be able to assist with services supporting a required certification in certain markets.

The availability of regional auditors can be a significant factor to keep auditing costs under control, as travel cost can be contained. Some of the smaller Notified Bodies work with local partners. Most of the time, this is working smoothly. But from a customer's point of view he has to work with two (or more) separate entities, which may make communication more difficult and may result in further challenges and problems.

When selecting a Notified Body, a manufacturer should also take into consideration the strategy in staffing of this organization. A designated project manager at the

Notified Body can help making the communication process more transparent and easy. This project manager's task is to see to it, that all agreed services are carried out and documented. He also facilitates the interaction between the departments and experts of the Notified Body.

The availability of dedicated assessors for the review of technical files and design dossiers can be the determining factor for whether or not an agreed timeline can be met. Many Notified Bodies do not have dedicated resources for the assessment of files, which frequently leads to situations where the same person who is scheduled to perform the assessment of the product will spend some weeks on an auditing tour. Such a situation can result in delaying the issuing of a desired certificate.

It is equally important that an assessor from time to time has the opportunity to accompany a lead auditor at an audit of a company for which he or she is assessing technical documentation. Only then can the assessor gain sufficient insight into the systems of the company and a solid understanding of how procedures described in the documentation are transferred into reality. These audits can also help to increase the level of trust of the Notified Body in the company and its systems.

When dealing with a Notified Body, a certain level of bureaucracy is inevitable. When selecting a Notified Body, the manufacturer should consider the internal organization and the approval procedures of the Notified Body.
Frequently delays in issuing certificates are caused by the fact that a central certification board has to meet and approve certificates after the assessment has been carried out. These boards do not meet frequently and this can cause delays. To avoid these delays, some Notified Bodies have a so-called "certification manager".

### 11.6. Summary of the Selection Criteria of a Notified Body
The following overview gives examples how a company can decide on the selection criteria of a Notified Body:

|  | NB 1 | NB 2 | NB 3 | NB 4 | NB 5 |
|---|---|---|---|---|---|
| **Experience with the relevant product range** | IOLs; solutions; | Ophthalmo- logical devices; implantable devices; active MDs; dispos- able MDs; | Ophthalmo- logical devices; laser; implant- able devices; active MDs; solutions; disposable MDs; | Ophthalmo- logical devices; eye care solutions, laser, active devices, IVDs, implantable devices | Laser; active diagnostic devices; implantable devices; active MDs; dispos- able MDs |
| **Other customers** | Company A | Company C, D | Company A, B, E, F | Company G | Company A |
| **Regional ISO & MDD auditors available** | No | Yes | Yes | Yes | (No) |
| **One central project manager** | Yes | Yes | Yes | Yes | No |
| **Dedicated dossier assessors available** | Yes | No | 20 in EU 8 in US | Yes | No |
| **Central certification board** | Yes | Yes | Certification manager | Yes | Yes |
| **Formal "fast track" appro- val option** | Yes | Yes | Yes | Yes | No |
| **Fixed time- lines for assessments** | Yes 3 months for significant changes 1 month for non-substantial changes | No | Yes Max. 8 weeks for first review 3 weeks to issue certifi- cate | Yes 3 months for significant changes or class II dossiers 1-2 months for non-substantial changes and class IIa/b files | No |

134

| | NB 1 | NB 2 | NB 3 | NB 4 | NB 5 |
|---|---|---|---|---|---|
| **Formal Project plan** | No | No | Yes | Yes | No |
| **Training offers** | Q-System | Management system; standards; Q-System | Q-System; MDD; Standards; Clinical trials; | Q-System; MDD; Standards; | Q-System; Standards; mainly technical training |
| **Clear definition of changes that require pre-approval** | +- | +- | Yes | +- | +- |
| **Approach: management review vs. process maps** | | | Management review only at CE "owner site". At manufac-turing site focus on process flow | Management review only at CE "owner site". At manufac-turing site focus on process flow | |
| **Sharing "best practice"** | No | No | Yes | Yes | No |
| **"Matrix certification" possibility** | No | No | Yes | No | No |
| **"ease" of adding new codes to existing families** | Update of certificate and DOC required. | Update of certificate and DOC required. | Very easy as certificates just mention pro-duct families. Product codes can be added to DOC. | Very easy as certificates just mention pro-duct families. Product codes can be added to DOC. | Very easy as certificates just mention pro-duct families. Product codes can be added to DOC. |
| **Available experience with move of certificates from NB to NB** | Yes | Yes | Yes | Yes | ? |

| | NB 1 | NB 2 | NB 3 | NB 4 | NB 5 |
|---|---|---|---|---|---|
| **Well established procedure to move from NB to NB** | Yes | Yes | Yes | Yes | ? |
| **"customer orientation"** | Act more like an authority | Strong | Very strong | Very strong | ? |
| **"Reputation" in MD industry** | Under very tight control of Irish Board of Medicines | Under tight control of UK MHRA; | Very flexible and pragmatic | Flexible & reliable | Only known to electrical equipment manufacturers |
| **Level of "control" by national MOH** | High | High | Low | Medium | |
| **"influence level" in NB working group on EU level** | Low | Significant | Significant | Medium | |

Table 11/1: Overview of the selection criteria for a Notified Body (Notified Body = NB; Medical Device = MD)

**Exercise:**

Go to the homepage of the European Union and look for the list of Notified Bodies and choose the selection criteria regarding your medical devices

## 11.7. Exchanging a Notified Body

A manufacturer has the option to select any Notified Body within the EU, as long as this organization is accredited to perform a conformity assessment for a particular product. This flexibility in the selection of a Notified Body also includes the option to exchange one Notified Body for another one. The already issued CE-certificates are then transferred from the former to the new Notified Body.

The business relationship between a manufacturer and a Notified Body should ideally be a long-term relationship. Even though the requirements for placing medical devices on the European market are the same for all manufacturers and Notified Bodies, there are differences in interpretation of requirements and individual preferences regarding the format of documentation, frequency and content of notifications and other formal and practical aspects. Once the business relationship between a manufacturer and "his" Notified Body is established all the mentioned aspects can foster the ongoing cooperation.

In today's world of acquisitions and mergers, however, a company might profit from limiting the number of Notified Bodies with which it has a business relationship. This helps limit the number of "internal" formal procedural requirements and audits that have to be performed on an annual basis by a Notified Body. It also assures more consistency of opinions and interpretation of requirements of laws, directives and standards. Moreover, limiting the number of Notified Bodies can help a manufacturer increase his importance as a customer, and thus increase the attention and service level from the Notified Body.

Many of the Notified Bodies have defined procedures on how to transfer CE certificates from one Notified Body to another. A transfer usually involves an evaluation of the existing CE certificate and the most recent assessment report of the future Notified Body. Additional audits might be required which usually will be conducted at the scheduled date of the next audit. As a consequence of a transfer to a new Notified Body, a revised declaration of conformity will have to be issued and the product labelling will have to be revised to reflect the new Notified Body's reference number.

Especially after acquisitions and mergers, consideration should be given to centralizing ownership of CE certificates at the legal entity that is in charge of compliance of products with global regulations. This is another helpful step to reduce the number of required audits at various company sites. This facilitates consolidation of processes and internal requirements.

11.8. Summary

Manufacturers of medical devices planning to promote their products within the EU should be well aware of their needs and the knowhow of a Notified Body before committing to the Notified Body on a contractual basis. The most important criteria is the accreditation of the Notified Body (for assessing existing and future products of a company), followed by its experience in this area.

Further criteria are the willingness of the Notified Body to share its knowledge in this specialized field with the manufacturer, its size, global presence as well as dedicated project managers and assessors. The Notified Body must have clearly defined requirements concerning product modifications.

The careful and sophisticated selection of a Notified Body saves time and money and increases the chances of a company of successfully entering the market.

## 11.9. Test Your Knowledge

| |
|---|
| **Q1:**  What activities are usually performed by a Notified Body? |
| 1. Design examination |
| 2. Production and Product Quality Assurance |
| 3. Composition of the technical documentation |
| 4. Type examination |
| **A1:**  1, 2 and 4 |

| |
|---|
| **Q2:**  What criteria have to be considered when selecting a Notified Body? |
| 1. Accreditation of certifying various MD classes |
| 2. Experience with the relevant product line |
| 3. Size and global presence |
| 4. Experience with specialized products |
| **A2:**  1 – 4 |

## 11.10. References

- The Medical Device Directive 93/42/EEC dated June 14, 1993
- The active Implantable Medical Device Directive 90/385/EEC dated June 20, 1990
- The In Vitro Diagnostic Medical Device Directive 98/79/EEC dated October 27, 1998

- http://ec.europa.eu/enterprise/newapproach/nando/index.cfm?-fuseaction=directive.notifiedbody&dir_id=13
- Notified Body Operations Group, NBOG, www.nbog.eu

Criteria for the Selection of a Notified Body

## Chapter 12: Usability

*Dr. Carsten Rupprath, Myriam Becker*

12.1. Learning Objective

The term "usability" can be referred to all kinds of commodities – and also to medical devices. This chapter will give you a basic understanding of the complex usability process. You will be able to become acquainted with and understand usability measures regarding the design of medical devices.

12.2. Introduction

The more medical devices are distributed in the markets, the more avoiding risks and dangers which occurred by using medical devices became an important topic. One aspect is avoiding use errors via adequate usability of medical devices. A manufacturer of medical devices has to guarantee the usability of his products. The objective of the usability engineering process is to assess and mitigate usability risks. Without considering the usability of a medical device already in the design process, there might be products in the end that are not intuitive and difficult to use. The design of user interfaces asks for special abilities in order to guarantee an adequate and safe usability of medical devices that are getting more and more complicated and are also used by laymen. This task clearly differentiates from the pure technical implementation of a user interface. By focusing on the usability already at the beginning of the design process, usability problems can be identified and minimized at an early stage.

12.3. Application of the Relevant Standard for Medical Devices: IEC 62366-1

A manufacturer can – to a certain extent – influence incorrect use of his product. Therefore, a risk management process (IEC 62366-1) can help keeping the usability in mind when designing a product. The IEC 62366-1 standard defines a process for a manufacturer to analyze, specify, design, verify and validate usability, as it relates to safety of a medical device. This usability engineering process assesses and mitigates risks caused by usability problems associated with correct use and use errors (i. e. normal use). It can be used to identify risks but does not assess or mitigate risks associated with abnormal use.

According to IEC 62366-1, the manufacturer has to implement and document a usability-oriented design process in order to provide safety to patients and users. Therefore, IEC 62366 asks the manufacturer of medical devices to perform a process for analysis, design, verification and validation of the usability in case these affect the safety of the medical device. The usability process should look at the correct use as well as the abnormal use. The usability process according to DIN EN 62366 can be incorporated in the quality management system of a manufacturer. The design process should define user-interactions regarding the medical device for:

1. Transport
2. Storage
3. Installation
4. Operation
5. Maintenance and repair
6. Disposal

The following text in italics refers to the outline of IEC 62366 (IEC 62366-1, ed. 1.0) which deviates from the currently valid standard (IEC 62366:2007).

*Risk control – to a certain extent – relates to usability. In Section 4.1.2 of IEC 62366 and in ISO 14971:2007 the reduction of usability risks is addressed and that the manufacturer considers one or more of the following techniques:*

1. *Inherent safety by design (f. e. user interface that has safeguards not to accept user input that is out of range)*
2. *Protective measures in the medical device itself or in the manufacturing process (f. e. alarm systems)*
3. *Information for safety (f. e. warnings in the instruction for use)*

*If information for safety is used as risk control measure, the manufacturer shall subject this information to the usability process to determine that the information is understandable to users of the intended user profiles in the context of the intended use environment (4.1.3). Disregarding such information for safety shall be considered an intentional act (f. e. sabotage or violative, contraindicated or reckless use) and is therefore beyond any further practicable means of user interface-related risk control by the manufacturer.*

## 12.4. Conformity with IEC 62366

By complying with this standard and meeting the usability validation criteria, residual risks associated with usability of a medical device are presumed to be acceptable for being used in the risk management process according to ISO 14971 (IEC 62366: 4.1.2). The results of the usability-oriented design process (Usability Engineering/Human Factors Engineering – UE/HFE) have to be documented in the UE/HFE-file as defined in section 4.2 (IEC 62366).

*The effort used to fulfill the UE/HFE process may vary based on the size and complexity of the user interface, the severity of the harm associated with the usability of the medical device, or the use specifications. In the case of a modification of an existing medical device design that has been subjected to the UE/HFE process, the effort used to fulfill the process steps in the UE/HFE process may be scaled based on the extent of the modification and the severity of the harm associated with the usability of the user interface (IEC 62366: 4.3).*

In the UE/HFE process the manufacturer has to prepare the use specifications *(IEC 62366:* 5.1), which takes into account the intended medical indication, intended patient population, intended part of the body or type of tissue applied to or interacted with, intended user profile, use environment and operating principle.

*In a next step of this process, the hazard-related use scenarios have to be identified (IEC 62366: 5.2). The manufacturer shall determine the primary operating functions and the inputs to the primary operating functions shall include the functions related to the safety of the medical device (IEC 62366: 5.2.1). In a following step the manufacturer shall analyze ways in which the user interface could contribute to use errors and use failures.*

Based on this analysis, an identification of the user interface characteristics related to safety (part of a risk analysis) that focuses on usability shall be performed according to ISO 14971:2007, 4.2 (IEC 62366: 5.2.2). The manufacturer shall identify known or foreseeable hazards related to the usability according to ISO 14971:2007, 4.3. Reasonable foreseeable hazard-related use scenarios associated with the medical device also shall be identified (IEC 62366: 5.2.3).

*The manufacturer shall describe hazard-related use scenarios. He may prioritize the hazard-related use based on the severity of the potential harm. When the medical device has a large number of hazard-related use scenarios, it is appropriate for the manufacturer to focus its attention and effort on the user interface elements that could have the most impact on the users´ interactions with the medical device and the related risk (IEC 62366: 5.3 and 5.4).*

*The manufacturer shall prepare and maintain a user interface specification (IEC 62366: 5.5). The user interface specification shall consider the use specification, the known or foreseeable use errors and use failures and the hazard-related use scenarios. The manufacturer shall prepare and maintain a user interface evaluation plan. This plan shall contain testable requirements for determining whether primary operating functions are recognizable, understandable and operable by the user. It should also contain usability tests which document the involvement of representative intended users and the individual test environment based on the use specifications (IEC 62366: 5.6).*

---

**Exercise**

Think about products of your every-day life where considering usability in the design process may have been of advantage. This underlines the importance of usability with medical devices because applying them the right way may well be a matter of life or death.

---

The user interface evaluation plan consists of a formative evaluation (IEC 62366: 5.6.2) as well as a summative evaluation (IEC 62366: 5.6.3). The manufacturer shall design and implement the user interface as described in the user interface specification utilizing, as appropriate, UE/HFE methods and techniques, including formative evaluation (IEC 62366, 5.7).

*The manufacturer shall perform a summative evaluation of the primary operating functions on the final implemented or production equivalent user interface according to the user interface evaluation plan (IEC 62366: 5.8). The results of any summative evaluation shall be analysed to determine whether the acceptance criteria documented in the user interface evaluation plan were met. Moreover, the results of any summative evaluation shall be analysed to identify the root cause of any user errors, use failures or other use problems that occurred. The root causes should be determined based on observations of user performance and subjective comments from the user related to that performance. The root causes should then be correlated with the risk analysis to identify the severity of the harm associated with the use errors, use failures or other problems. The need to implement and the feasibility of implementing additional risk control measures should then be determined. Further user interface design and implementation activities shall be performed (if possible). If further user interface improvement is not practicable, the manufacturer has to document the residual risk as well as perform a residual risk analysis according to ISO 14971:2007.*

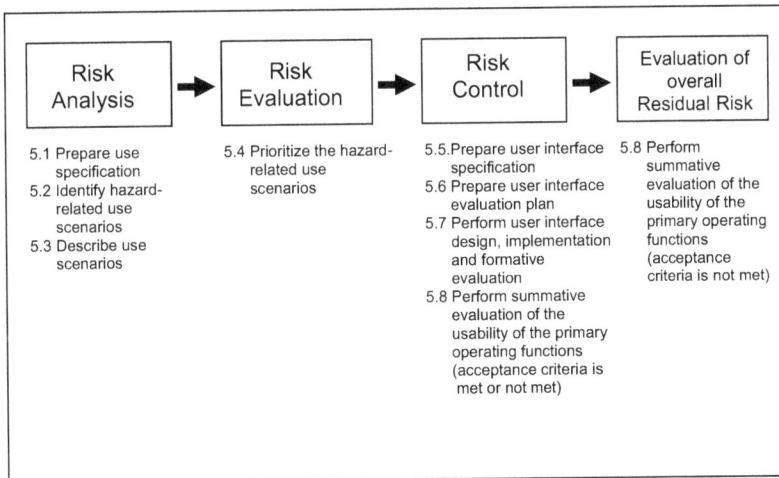

| Risk Analysis | Risk Evaluation | Risk Control | Evaluation of overall Residual Risk |
|---|---|---|---|
| 5.1 Prepare use specification<br>5.2 Identify hazard-related use scenarios<br>5.3 Describe use scenarios | 5.4 Prioritize the hazard-related use scenarios | 5.5.Prepare user interface specification<br>5.6 Prepare user interface evaluation plan<br>5.7 Perform user interface design, implementation and formative evaluation<br>5.8 Perform summative evaluation of the usability of the primary operating functions (acceptance criteria is met or not met) | 5.8 Perform summative evaluation of the usability of the primary operating functions (acceptance criteria is not met) |

Fig. 12/1: Usability Engineering/Human Factor Engineering Process (UE/HFE-Process) according to IEC 62366-1

The accompanying document (e.g. IFU, technical description, installation manual) shall include a summary of the use specifications, a concise description of the medical device and where relevant to its use: operating principle, technical characteristics, performance characteristics and intended user profile (IEC 62366: 6). If training on the specific medical device is required for the safe and effective use of a primary operating function by the intended user, the manufacturer must enable the user to get this training by providing the material necessary for the training and by making the training available (IEC 62366: 7). If such a training is required, the accompanying document shall describe the available training options and shall include the suggested duration and frequency of such training.

### 12.5. DIN EN ISO 9241

Also mentioned in regards to usability should be DIN EN ISO 9241. This standard covers the ergonomics of human-computer interaction that means the workplace, hardware and software.

In section 11 of this standard, the requirements for the usability are addressed. The usability of software depends on the intended use context (user, task, equipment as well as physical and social environments). There are three main criteria for the usability of software:

- Effectiveness (task completion)
- Efficiency (task in time)
- Satisfaction (responded by user in term of experience)

Section 110 of this standard deals with general ergonomic principles that apply to the design of dialogues between humans and information (interactive) systems. This refers to both – software and hardware. User interfaces of interactive systems – like websites or software – should be easy to use by the user. The term "user interface" is defined as "all parts of an interactive system that provide information or operating elements which are necessary for the user to complete a certain task via the interactive system".

There are seven principles for dialogue design listed in section 110:

- Suitability for the task
- Self-descriptiveness
- Controllability
- Conformity with user expectations
- Error tolerance
- Suitability for individualization
- Suitability for learning

## 12.6. Glossary

In this chapter you will find a variety of terms that are linked to the topic "usability" with short explanations. The definitions in italics are taken from ISO 62366 standard (among others).

- Abnormal use
  Intentional act or intentional omission of an act of the responsible organization or user of a medical device as an act that is beyond any further practicable means of risk control by the manufacturer.

- Normal use
  Operation, including routine inspection and adjustment by any user, and stand-by according to the instructions for use, or in accordance with generally accepted practice for those medical devices provided without instructions for use.

- Primary operating function
  User interface function for a user task or series of tasks that is related to the safety of the medical device.

- Usability
  Characteristic of the user interface that facilitates use and thereby establishes effectiveness, efficiency, ease of user learning and user satisfaction in a specified context of use.

- Usability Engineering/Human Factors Engineering (UE/HFE)
  Application of knowledge about human behaviour, abilities, limitations and other characteristics related to the design of tools, machines, equipment, devices, systems, tasks, jobs, and environments to achieve adequate usability and thereby acceptable risk related to use.

- UE/HFE-file
  Set of records and other documents that are produced by the UE/HFE process.

- Use error
  Act or omission of an act that leads to a different result than intended by the manufacturer or expected by the user.

- User interface
  Means by which the user and the medical device interact.

- Validation

Confirmation, through the provision of objective evidence, that the requirements for a specific intended use or application have been fulfilled.

On the website ([www.causause.de/wissen/usability-glossar.html](www.causause.de/wissen/usability-glossar.html)) a glossary regarding usability is provided with explanations on usability and user experience. This list covers a lot of terms – from accessibility, discount usability engineering, joy of use, prototyping, software ergonomics, usability lab to web usability.

## 12.7. Summary

The majority of manufacturers do not place medical devices with low usability on the market. The consequences would be too harmful: a huge number of user complaints, or even the death of a patient, high numbers of services supplied by the manufacturer, a low number of applications of the device, an expensive and time-consuming re-design of the product and consequently a decline in image and earnings.

The last outline of the standard that describes the usability for medical devices was published in November 2012. It requires manufacturers of medical devices to conduct a systematic usability engineering process that results in a reduction of use errors via adequate medical device usability. This will result in a higher product satisfaction by the users.

## 12.8. Test Your Knowledge

| | |
|---|---|
| **Q1:** | Which international standard takes care of the topic "usability of medical devices"? |
| **A1:** | IEC 62366:2007 |

| | |
|---|---|
| **Q2:** | Name at least four out of seven dialogue principles defined in DIN EN ISO 9241-110? |
| **A2:** | Suitability for the task, self-descriptiveness, controllability, conformity with user expectations, error tolerance, suitability for individualization and suitability for learning |

## 12.9. References

- DIN EN 62366 Medical devices usability
- IEC 62366-1 Ed. 1.0 - Committee Draft of IEC 62366-1: Medical devices – Part 1: Application of usability engineering to medical devices; date of circulation: 2012-11-23
- ISO 14971 – Medical Devices: Risk Management
- DIN EN ISO 9241 Ergonomics of human-computer interaction, section 11, Usability requirements, section 110: dialogue principles
- www.causause.de/wissen/usability-glossar.html

Usability

## Chapter 13: Quality Management

*Dr. Carsten Rupprath*

13.1. Learning Objective

In this chapter you will become acquainted with the basics of quality management and you will understand how risk management, design control and the change management interact in order to achieve the highest product quality with batch to batch consistency. Contents and requirements of EN ISO 13485 are described as well as a process-oriented approach and the steering and control of documents.

You will understand the terms "validation", "verification" and "auditing" as well as the role of the risk management within the overall management process. This chapter also explains how to cope with modifications of the medical device or of the quality management system.

13.2. Introduction

The requirements to medical device manufacturers regarding the quality of their products are high because all patients should be supplied with products of consistent high quality. Moreover, the manufacturers want to convince their customers with the high quality of their products and the consequences of defective goods can be life-threatening for patients. Quality is defined by all properties and features of a product that contribute to the compliance of given requirements (DIN 55350). The quality management system (QMS) describes and monitors all phases of the product realization process. Verification institutions like Notified Bodies audit and certify the QMS regularly.

13.3. EN ISO 13485 – The Basis of the QMS of Medical Devices

The implementation of a QMS according to EN ISO 13485 enables a manufacturer to design, produce and manufacture medical devices that are consistently complying with customer and regulatory requirements. The QMS can be used by an organization for the design and development, production, installation and servicing of a medical device, and the design, development and provision of related services (EN ISO 13485: 0.1). This standard can also be used by internal and external parties (including Notified Bodies) to assess the organization's ability to meet customers and regulatory requirements. The implementation of a QMS according to EN ISO 13485 is a system requirement in order to comply with legal and regulatory requirements.

The adoption of a QMS is a strategic decision of an organization that has influence on several parts of the organization (EN ISO 13485). EN ISO 13485 is a stand-alone standard but it is based on ISO 9001. Differences are the requirements for a continuous improvement process and customer satisfaction.

The requirements of MDD 93/42/EEC to a manufacturer of medical devices can be met to a certain degree by the manufacturer by complying with European harmonized standards. A European harmonized standard meets certain aspects of a European directive. Frequently, new harmonized standards of European directives are published in the Official Journal of the EU. If a manufacturer applies these harmonized standards he can be sure to be in compliance with the European directives.

In the MDD 93/42/EEC (Medical Device Directive) several conformity assessment procedures are described in order to ensure the conformity of the product with the directive. The manufacturer can choose a conformity assessment procedure but there are also limitations depending on the class of the medical device. The different procedures are described in the annexes of MDD 93/42/EEC. The often chosen conformity assessment procedure according to Annex II but also the procedures according to Annex V and VI assume that the manufacturer has already implemented a full quality assurance system according to EN ISO 13485. This should guarantee a consistent product quality and an improvement of the product in case that weaknesses in the design of a product become evident. The consistent quality of a medical device is of high importance for a low-risk application to the patient.

The standard EN ISO 13485 is based on a process approach to the implementation, development and improvement of the efficiency of a QMS. The application of a system of processes within an organization, together with the identification and interactions of these processes, and their management, can be referred to as the "process approach". The objective is to increase customer satisfaction by meeting customer requirements. Any activity that receives input and converts them to outputs can be considered as a process (EN ISO 13485: 0.2).

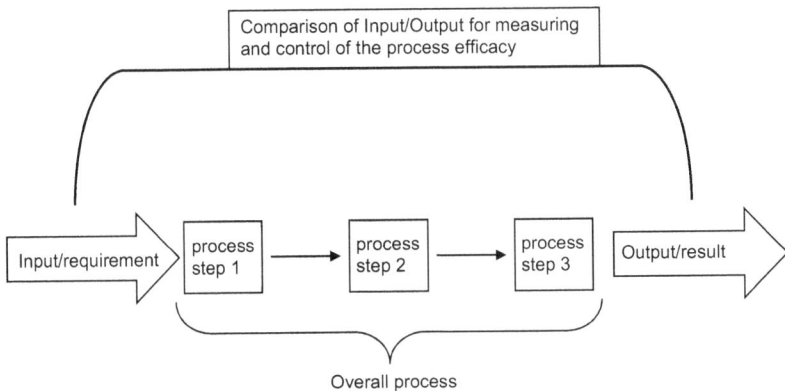

Figure 13/1: A process in a process approach

A process approach supports the implementation of a QMS in an organization. Any activity that receives input and converts them to outputs can be considered as a process. For an organization to function effectively, numerous connected and linked processes have to be identified and controlled. Often the output from one process directly forms the input to the next.

The application of a system of processes within an organization, together with the analysis of process results and their interactions, as well as their management, can be referred to as the "process approach" (Fig. 13/1).

The structure of processes can be demonstrated in a flow chart. The general objective of a QMS is the continuous improvement of the system by monitoring, steering, analysing and improving the processes.

13.3.1. Structure of EN ISO 13485

This standard consists of the following chapters:

1. Introduction (chapter 0)
2. Scope (chapter 1)
3. Normative references (chapter 2)
4. Terms and definitions (chapter 3)
5. Quality management system (chapter 4)
6. Management responsibilities (chapter 5)

7. Resource management (chapter 6)
8. Product realization (chapter 7)
9. Measurement, analysis and improvement (chapter 8)

Applying chapters 5-9 leads to the implementation of a quality management system. The processes start with the main processes of the organization and are developed from that level down to the level of single process steps via SOPs (Standard Operating Procedures) or working instructions. From every main process a number of lower level single processes with higher complexity and more details are derived. The main processes are established on the strategic level of the organization, e.g. production or sales.

Main processes are followed by business processes on the next, lower level. Business processes consist of work processes. Work processes describe the creation of an interrelated performance and can be carried out independently of each other. A work process consists of several work steps (or SOPs). SOPs cannot be subdivided any more as they consist of basic activities. The effectiveness of the described processes is monitored and assessed via measurable criteria.

---

**Exercise:**

Think about the kind of processes that exist.

---

13.3.2. Prerequisites for Getting Accredited According to EN ISO 13485

There are prerequisites for a QMS according to EN ISO 13485 that must be met before getting accredited by a Notified Body. The organization has to:

a) Identify the processes needed for the QMS and its application throughout the organization (chapter 4.1a)
b) Determine the sequence and interactions of these processes (chapter 4.1b)
c) Determine criteria and methods needed to ensure that both the operation and control of these processes are effective (chapter 4.1c)
d) Ensure the availability of resources and information necessary to support the operation and monitoring of these processes (chapter 4.1d)
e) Monitor, measure and analyse these processes (chapter 4.1e) and
f) Implement actions necessary to achieve planned results and maintain the effectiveness of these processes

The objective is the continuous improvement of the QMS. During the process of product realization the risk management (chapter 16 in the book) has to be taken into consideration in order to reduce/minimize the risk during production.

Guidelines for the application of EN ISO 13485 are included in the technical report according to ISO/TR 14969. Further guidelines were developed by the Global Harmonization Task Force (GHTF). The technical report according to ISO/TR 14969 offers guidance and supports the organization when designing, implementing and maintaining QMS according to EN ISO 13485. The legally binding EK-Med resolutions, as well as MEDDEV and NB-MED documents are published in the internet (EK-Med: www.zlg.de/medizinprodukte/dokumente/antworten-und-beschluesse-ek-med.html; MEDDEV:http://ec.europa.eu/health/medical-devices/documents/-guidelines/index_en.htm; NB-med: www.team-nb.org/).

Where an organization chooses to outsource any process that affects product conformity with requirements, the organization has to ensure control over such processes. Control of outsourced processes has to be identified within the QMS (EN ISO 13485: 4.1). Outsourcing can be carried out to another organization or to another business unit of an organization.

The standard EN ISO13485 has specific requirements on the documentation of the QMS.

### 13.3.3. Documentation Requirements

The quality management system documentation has to include:

a) Documented statements of a quality policy and quality objectives
b) A quality manual
c) Documented procedures
d) Documents needed by the organization to ensure effective planning, operation and control of its processes
e) Records (EN ISO 13485: 4.2.4)
f) Any other documentation specified by national or regional regulations.

The quality manual describes the scope of the QMS, the documented procedures established for the QMS as well as the interaction between the processes (EN ISO 13485: 4.2.2).

The quality manual has to outline the structure of the documentation used in the QMS. EN ISO 13485, chapter 4.2., also defines specific requirements to the technical documentation. The technical documentation generally is the documented proof (as result of the QMS) that demonstrates that the product complies with the essential requirements. Product specifications, quality requirements for the production as well as for the installation and servicing (if applicable) are part of it as well. The Summary Technical Documentation (STED) can be submitted to the Notified Body for assessment of conformity (except for medical devices of class I, see chapters 2 – 4 in the book).

The organization has to establish and maintain the technical documentation that defines the requirements to the QMS and the product specifications for every model of a medical device. The technical documentation has to include the whole production process and – if applicable – the installation and servicing (for more information see chapter 8 in the book).

### 13.3.4. Retention Period

Records have to be established and maintained to provide evidence of conformity to requirements and of the effective operation of the QMS. A documented procedure has to be established to define the controls needed for the identification, storage, protection, retrieval, retention time and disposal of the records.

The organization has to retain records for a period of time at least equivalent to the lifetime of the medical device as defined by the organisation but not less than 2 years from the date of product release or as specified by relevant regulatory requirements.

The regulatory requirements of MDD 93/42/EEC define a documents retention time of at least 5 years, and at least 15 years for implantable devices (from the date of last manufacturing of product). It is important to have the valid version of applicable documents at the location where they are needed, documents have to remain legible and readily identifiable in order to prevent the use of out-of-date documents.

13.3.5. Proving the Effectiveness of a QMS

The top management of an organization has to provide evidence of its commitment to develop, implement and maintain the effectiveness of a QMS.

a) Communicating to the organization the importance of meeting customer as well as statutory and regulatory requirements,

b) Establishing a quality policy

c) Ensuring that quality objectives are established,

d) Conducting management reviews

e) Ensuring the availability of resources (EN ISO 13485: 5.1).

Customer requirements have to be determined and met (EN ISO 13485: 5.2). Top management has to ensure that the quality policy is appropriate to the purpose of the organization and that it includes criteria which allow defining and evaluation of quality objectives. Moreover, the effectiveness of the QMS has to be maintained. The QMS has to be communicated and understood within the organization as well as to be reviewed for continuing suitability. Top management has to ensure that quality objectives are established at relevant functions and levels within the organization. The quality objectives have to be measurable and consistent with the quality policy (EN ISO 13485: 5.4.1).

Moreover, top management has to ensure that:

a) The planning of the QMS is carried out in order to meet the general requirements (EN ISO 13485: 4.1) as well as the quality objectives

b) The integrity of the QMS is maintained when changes to the QMS are planned and implemented (EN ISO 13485: 5.4.2).

Top management has to ensure that the responsibilities and authorities are defined, documented and communicated within the organization, has to establish the interrelation of all personnel who manage, perform and verify work affecting quality and finally has to ensure the independence and authority needed to perform these tasks (EN ISO 13485: 5.5.1).

Top management shall appoint a representative who shall have responsibility and authority to:

- Ensure that processes needed for the QMS are established, implemented and maintained
- Report to top management on the performance of the QMS and any need for improvement
- Ensure the promotion of awareness of regulatory and customer requirements throughout the organization (EN ISO 13485: 5.5.2).

### 13.3.6. Internal Communication and Management Review

Appropriate communication processes have to be established within the organization to ensure that communication takes place regarding the effectiveness of the QMS (EN ISO 13485: 5.5.3).

Top management has to review the organization's QMS at planned intervals in order to ensure its continuing suitability, adequacy and effectiveness. This review has to include assessing opportunities for improvement and the need for changes to the QMS including the quality policy and quality objectives.

The input to management review has to include information on:

1. Process performance and product conformity
2. Status of preventive and corrective actions
3. Customer feedback
4. Results of audits
5. Follow-up actions from previous management reviews
6. Changes that could affect the QMS
7. Recommendations for improvement
8. New and revised regulatory requirements

The management review has to define and document criteria for the acceptability of risks. These criteria should be based on standards, norms and Regulatory requirements. Reservations by participants of the management review and state of the art technical developments should be taken into consideration in the management review. Moreover, the management review has to take a look at the suitability of the risk management process at planned intervals thus ensuring the effectiveness as well as documenting implemented actions and decisions.

13.3.7. Resource Management

The provision of resources plays an important role in the QMS according to EN ISO 13485 (chapter 6.1.). The objective is to identify and provide resources in order to implement a QMS and maintain its effectiveness as well as to meet regulatory and customer requirements.

As far as human resources are concerned (EN ISO 13485: 6.2.), it is important to consider that personnel performing work affecting product quality has to be competent on the basis of appropriate education, training, skills and experience. The organization has to determine the necessary competence for personnel performing work affecting product quality, provide training or take other actions to satisfy these needs, evaluate the effectiveness of the actions taken, ensure that its personnel are aware of the relevance and importance of their activities and how they contribute to the achievement of the quality objectives and maintain appropriate records of education, training, skills and experience. An example for national training requirements is the "Medizinprodukteberater" (medical devices consultant) in Germany who is only allowed to perform his task after a specific training for obtaining sufficient knowledge regarding application of a specific medical device.

---

**Exercise:**

Think about methods that can be applied in order to assess the effectiveness of trainings.

---

As far as infrastructure (EN ISO 13485: 6.3) is concerned, the organization has to determine, provide and maintain the infrastructure needed to achieve conformity to product requirements. Infrastructure includes (as applicable) buildings, workspace and associated utilities, process equipment and supporting services (such as transport or communication). The organization has to define and ensure that documented requirements for infrastructure maintenance are met as this can influence product quality. Moreover, the organization has to determine and manage the work environment needed to achieve conformity to product requirements (EN ISO 13485: 6.4). In case that the contact between the personnel and the product or the work environment could adversely affect the quality of the product, the organization has to establish documented requirements for health, cleanliness and clothing of personnel (EN ISO 13485: 6.4). If work environment conditions can have an adverse effect on product quality, the organization has to establish documented requirements for the work environment conditions and documented procedures or work instructions to monitor and control these work environment conditions. Moreover, the organization has to ensure that all personnel who are required to work under special environmental conditions are appropriately trained or supervised by a trained person. If appropriate, special arrangements have to be established and documented for the control of contaminated or potentially contaminated products in order to prevent the contamination from spreading (to other products, to the work environment or personnel).

---

**Exercise:**

Think about two examples in which infrastructure or maintenance have an impact on product conformity of a medical device.

---

### 13.3.8. Product Realization

Chapter 7 of the EN ISO 13485 describes the product realization process. The organization has to plan and develop the processes needed for product realization (chapter 7.1). In the design phase the following properties are of high importance: 1. quality objectives and product requirements, 2. the need to establish processes and documents and to provide resources specific to the product, 3. required verification, validation, monitoring, inspection and test activities specific to the product and the criteria for product acceptance and 4. records to provide evidence that the realization process and resulting product comply with the requirements. For the risk management, requirements have to be defined for the whole product realization process and results from the risk management (chapter 15 in the book) have to be documented.

During product realization customer-related processes have to be considered. The organization has to determine requirements specified by the customer (including the requirements for delivery and post-delivery activities) and also requirements not stated by the customer but necessary for specified or intended use as well as statutory and regulatory requirements related to the product (EN ISO 13485: 7.2.1). The organization has to review the requirements related to a product and to conduct this review prior to the organization's commitment to supply a product (EN ISO 13485: 7.2.2). Moreover, the organization has to determine, define and implement effective arrangements for communicating with customers in relation to product information, enquiries, contracts or order handling, customer feedback (including customer complaints) as well as advisory notices (EN ISO 13485: 7.2.3). Advisory notices are further information of the organization on activities regarding the application of a medical device – after product release. This could be the modification of a medical device, the return of the medical device to the organization or the destruction of the medical device.

The QMS has also to be considered in the design phase. The organization has to plan and control the design and development of a product and establish documented procedures for this process (EN ISO 13485: 7.3.1) that consists of the design and development stages, the review, verification, validation and design transfer activities and the responsibilities and authorities for design and development. The design and development inputs include functional, performance and safety requirements, statutory and regulatory requirements, information derived from previous similar designs, other essential requirements to design & development as well as output of risk management. (EN ISO 13485: 7.3.2). Examples for regulatory requirements: medical devices MDD 93/42/EEC, IVDs 98/79/EC as well as WEEE 2002/96/EC or RoHS 2002/95/EC.

The outputs of design and development have to be provided in a form that enables verification against the design and development input. The outputs have to meet the input requirements for design & development, provide appropriate information for purchasing, production and for service provision, contain product acceptance criteria and specify the characteristics of the product that are essential for its safe and proper use (EN ISO 13485: 7.3.3).

The design & development review has to be performed in accordance with planned arrangements to evaluate the ability of the results of design & development to meet requirements and to identify any problems and corrective actions (EN ISO 13485: 7.3.4). A verification shall be performed in accordance with planned arrangements to ensure that the design & development outputs have met the design & development input requirements (EN ISO 13485: 7.3.5).

Design & development validation has to be performed in accordance with planned arrangements to ensure that the resulting product is capable of meeting the requirements for the specified application or intended use. Validation has to be completed prior to the delivery or implementation of the product (EN ISO 13485: 7.3.6). As part of design & development validation, the organization has to perform clinical evaluations and/or evaluation of performance of the medical device. The clinical evaluation is part of the technical documentation and has to be revised regularly using vigilance data and published literature. The transfer of design & development ensures that the design & development output is verified for the production before being used in the production phase.

Design & development changes have to be assessed, verified, validated and documented. These changes have to be approved before implementation and the review of design & development changes have to include evaluation of the effect of the changes on constituent parts and products already delivered (EN ISO 13485: 7.3.7). The organization has to evaluate and select suppliers based on their ability to supply products in accordance with the organization's requirements (EN ISO 13485: 7.4.1). Criteria for selection and evaluation of suppliers have to be established by the organization. Records of the results of the evaluations have to be established as well as inspections for ensuring that the purchased products meet specified purchase requirements. Moreover, the organization has to state the intended verification arrangements (EN ISO 13485: 7.4.3).

### 13.3.9. Production and Service Provision

The production and service provision also has to take place under controlled conditions that are defined in advance (EN ISO 13485: 7.5.1). Controlled conditions include the availability of documented procedures, of documented requirements, reference measurement procedures and work instructions. Moreover, the availability and use of monitoring and measuring devices as well as suitable equipment has to be guaranteed. The implementation of release, of delivery and of post-delivery activities, of defined operations for labelling and packaging have to be implemented by the manufacturer. The organization has to establish and maintain a record for each batch of medical devices to provide traceability. The batch record has to be verified and approved. Moreover, for sterile products, the organization has to maintain records of the process parameters for the sterilization process that was used for each batch (EN ISO 13485: 7.5.1.3).

**Exercise:**
Please define why traceability of products has to be guaranteed.

The organization has to validate any processes for production and service provision where the resulting output cannot be verified by subsequent monitoring and measurement. Validation has to demonstrate the ability of these processes to achieve planned results (EN ISO 13485: 7.5.2). The organization has to establish arrangements for these processes including defined criteria for review and approval of processes, approval of equipment and qualification of personnel, use of specific methods and procedures, requirements for records and revalidation. Computer software for production and service provision and measurement processes also has to be validated.

There are special requirements for sterile medical devices. The organization has to establish documented procedures for the validation of sterilization processes. The results of the sterilization validation have to be documented and recorded. Sterilization processes have to be validated prior to initial use (EN ISO 13485: 7.5.2.2). The organization has to identify a product by suitable means throughout product realization and establish documented procedures for such product identification (EN ISO 13485: 7.5.3.1). Also documented procedures for traceability have to be established (EN ISO 13485: 7.5.3.2). For implantable medical devices and active implantable medical devices there are special requirements for traceability (for details review EN ISO 13485: 7.5.3.2).

The organization has to exercise care with customer property while it is under the organization's control or being used by the organization (EN ISO 13485: 7.5.4). The preservation of the product during internal processing and delivery to the intended destination has to be guaranteed by the organization (EN ISO 13485: 7.5.5). This preservation includes identification, handling, packaging, storage and protection of the product.

---

**Exercise:**

Think about the reasons for validating sterilization procedures prior to use during the production of a medical device.

---

The organization has to establish documented procedures for the control of a product requiring special storage conditions.

The organization has to determine the monitoring and measurement to be undertaken and the monitoring and measurement devices need to provide evidence of conformity of the product to determined requirements (EN ISO 13485: 7.6). Moreover, the organization has to establish documented procedures to ensure that monitoring and measurement can be carried out in a manner that is consistent with the monitoring and measurement requirements.

The organization has to plan and implement the monitoring, measurement, analysis and improvement processes required:
1. To demonstrate conformity of the product
2. To ensure conformity of the QMS and
3. To maintain the effectiveness of the QMS (EN ISO 13485: 8.1).

13.3.10. Measurement, Analysis and Improvement

The organization has to implement a feedback system in order to collect information (e. g. customer complaints) and whether customer requirements are met by the organization (EN ISO 13485: 8.2.1). This feedback system can provide early warning of quality problems and result in input to corrective and preventive action processes. Post-market surveillance can offer input from many different sources for corrective measures. The following experience can result in corrective measures: long-term complications, reliability, performance problems, information from the risk management, change management or production problems. Also to be considered are: feedback to the instructions for use, customer satisfaction, usability, necessity for customer trainings, notifications to the authorities or knowledge on misuse (chapter 20 in the book).

The organization has to conduct internal audits at planned intervals to determine whether the QMS conforms to the planned arrangements, to the requirements of this international standard and to the QMS requirements established by the organization and to determine that the QMS is effectively implemented and maintained (EN ISO 13485: 8.2.2). The audit criteria, scope, frequency and methods have to be defined by the organization. The organization has to apply suitable methods for monitoring and measurement of the processes of the QMS in order to demonstrate the ability of the processes to achieve the planned results (product conformity). When planned results are not achieved, corrective action has to be taken to ensure product conformity (EN ISO 13485: 8.2.3).

The organization has to monitor and measure the product characteristics in order to verify product conformity (EN ISO 13485: 8.2.4). This has to be carried out at appropriate stages of the product realization process in accordance with the planned arrangements and documented procedures. Evidence of conformity with the acceptance criteria has to be maintained and documented.
Furthermore, the organization has to ensure that products that do not conform to product requirements are identified and controlled to prevent its unintended use or delivery (EN ISO 13485: 8.3). The controls and related responsibilities and authorities for dealing with nonconforming products has to be defined in a documented procedure and the organization has to take actions to eliminate the detected nonconformity. How the organization copes with nonconforming products has to be documented. This also includes acceptance of the product under concession. If a nonconforming product undergoes corrective measures, it has to be assessed again for conformity. If a nonconforming product is identified after its release, the organization has to undertake measures depending on the possible consequences.

---

**Exercise:**

Think about what could be minimal and maximal consequences of a nonconforming product for an organization. Go to the internet for information on "recalls".

---

## 13.3.11. Analysis of Data and Improvement

The organization has to establish documented procedures to determine, collect and analyse appropriate data to demonstrate the suitability and effectiveness of the QMS and to evaluate if improvement of the effectiveness of the QMS can be achieved (EN ISO 13485: 8.4). This includes data generated as a result of monitoring and measurement and from other relevant sources. The analysis of the data has to provide information relating to conformity with product requirements, characteristics and trends of process and product, feedback from customer and suppliers.

The organization has to put special focus to the maintenance and the effectiveness of the QMS (EN ISO 13485: 8.5). Quality policy, quality objectives audit results, data analysis, corrective measures and the management review have to be used to achieve this objective.

Records of all customer complaint investigations have to be maintained. If investigation determines that activities outside the organization contributed to the customer complaint, relevant information has to be exchanged between the organizations involved. If any customer complaint is not followed by corrective and/or preventive action, the reason has to be authorised and recorded. If national or regional regulations require notification of adverse events that meet specified reporting criteria, the organization has to establish documented procedures to such notification to regulatory authorities.

Corrective measures have to be carried out according to documented procedures. First of all, the nonconformity has to be identified, the cause to be assessed and appropriate actions to be carried out in order to prevent recurrence. The corrective measures are then assessed due to their effectiveness (EN ISO 13485: 8.5.2). All measures and results of the investigations have to be documented. Preventive actions should be implemented in order to eliminate the causes of potential nonconformities to prevent their occurrence (EN ISO 13485: 8.5.3). A documented procedure has to be established to define requirements for determining potential nonconformities and their causes, evaluating the need for action to prevent occurrence of nonconformities, determining and implementing action needed, recording of the results of any investigations and of actions taken and reviewing preventive action taken and its effectiveness.

13.4. Change Control

Modifications of medical devices or of the QMS of an organization occur regularly. The lifecycle of a medical device is – compared to a medicinal product – relatively short. Moreover, modifications to the QMS that lead to its improvement is intended and an objective of the QMS. It is important to know which modifications are reportable and what consequences these modifications could trigger. Changes of the product or the QMS are often a result of changes in design & development, in production, in the promotion of the product, in the regulatory environment or due to clinical results, economic changes or changes in the markets. Guidance regarding the assessment of a reporting obligation is provided by NB-MED/2.5.2/Rec2 (www.meddev.info).

Ideally, every organization has established a documented process in its QMS regarding change review (change control) and reporting obligation. The change control will be conducted by representatives of various functional departments (R&D, Regulatory Affairs, Sales, Quality Assurance, Marketing, Production, ...).

Reportable changes can then be divided in changes of the design, of the product or of the QMS. Also changes can be defined as significant/substantial or non-significant/non-substantial. The definition of significant changes refers to changes to the product, whereas substantial changes refer to changes to the QMS or to plans of changing of the product range. Product changes are changes that affect the compliance to the essential requirements of MDD 93/42/EEC and/or concern the requirements of the intended use.

Changes to the QMS are changes that may affect the conformity of the medical device with the essential requirements of the MDD 93/42/EEC.

The following changes to the QMS often occur:

1. Changes of the technology
2. Changes of the product family
3. Changes that affect the conformity with the essential requirements
4. Changes that affect the conformity with the harmonized standards
5. Possibly the following changes: validation, sterilization, used chemicals/ materials, supplier, organization structure and verification

The Notified Body has to be notified on significant and substantial changes. The change submission to the Notified Body has to include a description of the change as well as a comparison with the CE-marked variant. Moreover, the change has to be justified. If it is a change in product or design, the relevance in regard to the product conformity with the essential requirements has to be assessed and justified. The reporting obligation to the Notified Body vary according to the annexes of the MDD 93/42/EEC that are referred to for certification.

When certifying in accordance to Annex II, chapter 3.4. (that requires the implementation of a full QMS according to EN ISO 13485), every planned substantial change of the QMS or plans to change the product range have to be reported. Changes to the certified design of class III products which are approved in accordance to Annex II, chapter 4.4. need the approval of the Notified Body if the intended use or the essential requirements are affected.

Before implementing a change, the organization has to carry out a risk assessment of the planned change. In the course of the risk assessment, it can be decided on the reporting obligation and the NB authorization. In a next step, the technical documentation should be updated before any change is implemented. Depending on the change, the risk management documents, the clinical assessment, the list of the essential requirements, the list of the applied standards and the labelling/packaging has to be revised as well. Finally, the QMS and SOPs have to be updated regarding relevant information about the change.

The risk assessment of a change is of high importance for an organization because non-acceptable risks and preventable risks should be avoided. First of all, the probability of occurrence of an incident and the severity of the consequences has to be evaluated. Non-tolerable risks that outweigh the benefit of a change have to be avoided. The term "acceptable risks" means that the benefits of the change outweigh the neglectable risks. When reducing risks, it is important to consider all facts as well as continuously monitor the product in the market. A product can be considered safe, when it is free from unacceptable risks (chapter 16 in the book).

Substantial changes to the QMS are e. g.:

- The use of a new manufacturing technology
- The extension of a product family

- Changes that affect the product conformity
- Changes that affect the conformity of the QMS with harmonized standards or the relevant directive

### 13.5. Design Control

Changes during the product design phase have to be assessed whether they have influence on the components, the validated status of an established system, the process or the product. The main objective of the design change control is the identification and tracking of changes and to ensure that the problem is solved by the change (and no new problem is caused). Moreover, product documents have to be revised in regard to the new design.

Typical examples for changes in design control:

1. New product design
2. Changes of customer requirements
3. Changes of product requirements
4. Changes of the design output

When realizing a change during the product design phase, the change is first of all identified and then assessed by an expert group regarding feasibility, negative consequences and safety. Possible conflicts due to the change have to be excluded and an implementation plan has to be established. The next steps are creating the change control report and the change description with all planned differences. The change control report includes all changes within the QMS. The design history file (DHF) has to be updated as well because it contains the product history with all changes to the product. When assessing a change, it is important to consider whether the implementation affects the design plan, the customer requirements, the design validation, the design verification, the risk management file, the design transfer activities and the regulatory submission to the Notified Body. The sooner a change is implemented in the product design the smaller are the requirements resulting from this change. An inspection could be sufficient to fulfil the requirements for a design change which was early identified. If changes are implemented late in product design, many requirements have to be met again, e. g. the need for performing a new validation. Therefore, it is desirable for a manufacturer to implement changes as soon as possible in the design phase.

The documentation of the design control is of high importance because the requirements of the QMS have to be met as well as e. g. FDA requirements for design changes.

### 13.6. Summary

A QMS enables an organization to continuously assess and improve all processes which result in improving the quality of the products and the manufacturing processes.

The requirements of MDD 93/42/EEC regarding different conformity assessment procedures presuppose a full QMS and support the organization to meet customer and regulatory requirements resulting in customer satisfaction. The standard EN ISO 13485 describes the requirements to a QMS of a medical device organization and the standard is of high importance for manufacturers that distribute their products in Europe. Competent Authorities in other parts of the world like in Middle East, Africa, Asia-Pacific, Oceania and the CIS countries (non-EU east Europe) also accept and require certifications from organizations in accordance to EN ISO 13485 for the local product certification processes.

### 13.7. Test Your Knowledge

| | |
|---|---|
| **Q1:** | What is a process? |
| **A1:** | A process is an operation that starts with requirements (input) and finally results in outputs. |

| | |
|---|---|
| **Q2:** | How long does a manufacturer has to keep the technical documents for implantable medical devices according to the Medical Devices Directive? |
| **A2:** | 15 years. |

| | |
|---|---|
| **Q3:** | Why is it important for an organization to monitor the suppliers? |
| **A3:** | The supply of the organization with raw material, products or services can have a direct influence on product quality and product conformity. |

**Q4:** Why is the validation of manufacturing processes of high importance?

**A4:** A validation ensures that defined results e. g. in the production of a medical device can be achieved. This is of great significance for product conformity.

**Q5:** Is a manufacturer, when classifying a medical device to class I, required to establish a QMS in accordance to DIN EN 13485?

Is a QMS in accordance to ISO 9000 sufficient for a medical device manufacturer or is he required to take DIN EN 13485 as the basis for his QMS?

What are the differences or "consequences" if a medical device manufacturer "only" has a QMS based on ISO 9000?

**A5:** According to article 11.5 as well as Annex VII for class I products, the manufacturer only has to generate a technical documentation according to Annex VII, 3 as well as to issue a declaration of conformity. Please note that a clinical assessment has to be part of the technical documentation!

A quality management system is not a prerequisite for issuing a declaration of conformity or a CE-mark. Even for sterile medical devices or medical devices with a measurement function, establishing a QMS is not required. In this case, a Notified Body has to be involved that assesses the conformity of the sterility or measurement function. The assessment is conducted according to Annex V (Vs or Vm) with special focus on validation of the packaging, of the clean room and of the sterilization as well as all processes affecting the precision of the measurement function. In general, ISO 9000 and EN ISO 13485 are different regarding "customer satisfaction" and "continuous improvement". Both features that are included in EN ISO 13485 are not included in ISO 9000. When assessing the conformity with the medical device directive the compliance with the legal requirements plays a pivotal role (not customer needs) and "state of the art" (not continuous improvement). That means that a manufacturer of class I medical devices does not necessarily has to establish a QMS in accordance to EN ISO 13485.

**Q6:** Special case: Software as medical device. If a manufacturer intends to distribute software as medical device of class I, the manufacturer has to comply with the essential requirements of the MDD revision 2007/47/EC. In order to comply with chapter 12.1a, the manufacturer can apply the harmonized standard DIN EN 62304 (Medical Software). This standard refers to DIN EN 14971 as well as to DIN EN 13485 regarding an adequate risk management. Can one derive from this reference an obligation that this manufacturer has to have or should have a QMS in accordance to DIN EN 13485?

**A6:** The compliance with standards and harmonized standards is voluntary. From our point of view, a reference of a standard to another standard cannot result in an obligation. Maybe a manufacturer who does not meet the reference to e. g. EN ISO 13485, is not able to state full compliance to the standard. As this is not required, it should not be a problem.

### 13.8. References

- EN ISO 13485: QMSe requirements for regulatory purposes
- DIN 55350 Terms relating to quality assurance and statistic-terms for certification of quality assessment results
- MDD 93/42/EEC (http://eur-lex.europa.eu/LexUriServ/LexUriServ.do?uri=-CONSLEG:1993L0042:20071011:de:PDF
- Official Journal of the European Union (http://eur-lex.europa.eu/JOIndex.do?ihmlang=de)
- ISO/TR 14969 – Guidance for applying ISO 13485
- IVDs 98/79/EEC (http://eur-lex.europa.eu/LexUriServ/site/de/consleg/1998/L/01998L0079-20031120-de.pdf)
- EK-Med: www.zlg.de/medizinprodukte/dokumente/antworten-und-beschluesse-ek-med.html
- MEDDEV: http://ec.europa.eu/health/medical-devices/documents/guidelines/index_en.htm
- NB-MED: www.team-nb.org/
- www.meddev.info
- WEEE 2002/96/EG (http://eur-lex.europa.eu/LexUriServ/LexUriServ.do?uri=-OJ:L:2003:037:0024:0038:de:PDF)

- RoHS 2002/95 EG
  (http://eur-lex.europa.eu/LexUriServ/LexUriServ.do?uri=-
  OJ:L:2003:037:0019:0023:de:PDF)
- International Medical Device Regulators Forum: www.imdrf.org/
- Requirements to the QMS of a manufacturer (EN ISO 13485:2003/MPG),
  www.mdc-ce.de
- ICH Q9 (Quality Risk Management), The International Conference on Harmonisation of Technical Requirements for Registration of Pharmaceuticals for Human Use (ICH), www.ICH.org

## Chapter 14: Clinical Assessment of Medical Devices, "Literature Route"

*Dr. Stefan Menzl, Dr. Sibylle Scholtz*

14.1. Learning Objective

In this chapter you will learn about the structure and the various parts of a clinical assessment. You will be able to decide whether a clinical investigation has to be conducted or/and whether a literature search provides all needed, valid data.

14.2. Introduction

The clinical assessment of medical devices is an integral part of the scientific proof that a medical device complies with the essential requirements as stated in the MDD (93/42/EEC, Annex I, 6a) and its national implementations.

The clinical assessment provides important information that a medical device is safe in the clinical use and offers the (clinical) performance that the manufacturer claims in the product labelling and promotional material. The assessment of the clinical use of a medical device must be a perpetual process during the entire life cycle of a product. The first assessment takes place at the time the conformity assessment procedure is done. Further assessments of the clinical reliability and safety as well as the performance and therapeutic benefit of the medical device follow on a regular base. Current scientific state of the art should always be taken into account.

The input is generally provided by the risk analysis, risk management, literature and information on competitive, essentially similar technology. On the basis of this information, the manufacturer must decide whether or not newly gained data and elements of knowledge do have an influence on the benefit-risk assessment of the medical device and therefore have to be taken into consideration.

14.3. Significance of the Clinical Assessment of Medical Devices

In the clinical assessment the following elements have to be considered:

- The clinical performance of a product that is influenced by:
  - The applied technology
  - The biological safety
  - Usability
  - The intended use defined by the manufacturer
  - Product labelling

- o Warning labels
- o The experience of the user
- Residual risks that cannot be reduced any more by design modifications or warnings
- Post-market surveillance (PMS) that is based on the assessment of adverse events and other measures

During the recent revision in 2007, but also the currently debated revision of the Medical Device Directive, the clinical assessment was one of the parts that has been identified to need further improvement.

Other parts considered for improvement are: market surveillance, transparency, vigilance and consistence in choosing notified bodies as well as harmonization of the competence of the notified bodies (NBs).

### 14.4. Conducting the Clinical Assessment of Medical Devices

Directive 2007/47/EEC defines that a clinical assessment is mandatory for all medical devices (also non sterile class I products without measurement function), not only for high-risk devices and implants.

Generally speaking there are two ways for a manufacturer to obtain clinical data for a clinical assessment of his device:

- A clinical investigation to collect and assess clinical data regarding the use of the specific device
- Published data of similar devices for which equivalence to the device in question can be demonstrated (literature search)

Usually manufacturers use both ways. Before starting a work-extensive and expensive clinical investigation, a manufacturer first of all carries out a literature search in order to identify data from comparable devices. Only the data and information that cannot be retrieved by literature search is then addressed by conducting a clinical investigation with the specific device. This approach limits the extent of an investigation as well as the cost.

The objective of an investigation is to get data specific to the intended use of a medical device in order to prove its safety, performance and potential residual risks (93/42/EEC, Annex X, 1).

## 14.5. Recommendation for Conducting a Clinical Assessment

These steps should typically be followed when doing a clinical assessment for medical devices:

- Definition of databases to be searched
- Definition and description of the search strategy
- Documentation of the expected results
- Assessment of the available data regarding safety and performance
- Detailed analysis of the data regarding relevance and significance
- Summary and final conclusion

---

**Keep in mind!**

Every medical device has to be clinically assessed.

The clinical assessment is based on data derived from the product or a similar, equivalent device.

The clinical assessment has to be documented and updated regularly.

---

## 14.6. Detailed Information on the Literature Route

As already mentioned, a clinical assessment can include data of similar medical devices. The challenge for the manufacturer is to demonstrate that these products are indeed "similar" and "equivalent" to his specific medical device. Usually this data is derived from published clinical investigations, performed with similar, equivalent devices.

But also published and/or unpublished data on market experience with the device in question or a similar device for which equivalence to the device in question can be demonstrated – even beyond the study setting – can and should be taken into account.

The key questions to consider for selection and assessment of data should be as follows:

- Is the use of published data of similar devices an adequate strategy for my product (is there similar, equivalent technology available)?
- Which are the key characteristics of my product?
- Is available data derived from a previous model or from a similar device?
- In what sense is my new product similar to the described device?
- In what sense do these products differ?
- What data is available from post-market surveillance?
- What is the most appropriate search strategy to find relevant data?
- How can a reference to a similar, equivalent product be justified?
- Does the chosen literature support the indications and claimed performance features of my product?
- How is the ratio of literature supporting claimed product safety and characteristics versus literature questioning safety and characteristics?

When relying on published literature, it is important to assess the quality of this literature. It is of key importance whether the data was published in a peer-reviewed journal or in a so-called trade journal. Also the qualification of the author and his experience in the specific field is crucial. The quality of data is influenced by the qualification of the staff conducting the study as well as by the experience of the person who analyses the data.

Finally one has to decide at which point the literature search reached a state of saturation and can considered to be representative.

# Methodology        Claims

# Hypothesis

# Population        Validity

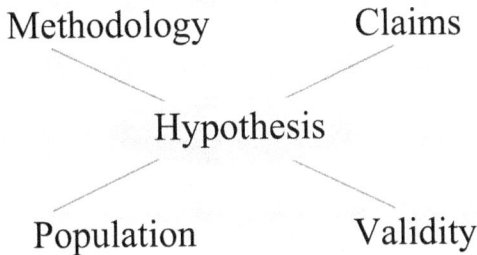

Fig. 14/1: Key criteria to assess the relevance of literature are the methodology used, the comparability of indications and performance characteristics and the comparability of study subjects

Published data can be taken from:

- Peer-reviewed journals
- Systematic reviews (Cochrane database)
- Unpublished data, so-called "data on file"

Published unsubstantiated opinions, random reports or reports lacking sufficient detail to permit scientific evaluation should not be taken into account.

14.7. Criteria for Assessing Equivalence/Comparability of Medical Devices

As far as equivalence/comparability of medical devices is concerned one needs to differentiate between:

- Technical equivalence/comparability
    - Comparable type of use
    - Comparable design
    - Comparable principles of operation
    - Comparable specifications and properties
- Biological equivalence/comparability
    - Use of same materials that are in contact with human tissue or body fluids
- Clinical equivalence/comparability
    - Comparable population
    - Comparable clinical use

- o Application at the same site of the body
- o Comparable performance features for the indication

When performing a literature search often the question arises when to stop the search. Two criteria may help to answer that question:

1. The search has shown an adequate number of results that contradict the intended performance claims for the product
2. A small modification of the search criteria does not result in a significantly higher number of results

It is also worthwhile to do a separate search per major indication or performance feature of a product.

## 14.8. Presentation of the Results

When presenting the results of a literature search, the following aspects should be highlighted:

- Chosen databases
- Search details per performance criterion
- Results per search detail incl. reference
- Relevance of the individual results
- Quality of the source
- Qualification of the author
- Criteria for saturation and evidence

Cave: the literature search has to be reproducible. A Notified Body has to get to the same results and conclusion when applying the same search strategy!

Therefore the search strategy has to be described. The major abstracts or full version of the texts should be part of the documentation. In any case, a list with all hits including the sources and relevance should be added.

Finally, the results of relevant sources should be summarized. There should be a representative ratio between positive and critical results. On this basis, a conclusion can be drawn regarding the benefit-risk ratio.

The results of the literature research are then combined with further results from other sources.

It is helpful to have the following parts of the technical documentation readily available to be considered in the overall-assessment:

- Intended use and defined product performance
- Specification of the product
- Preclinical assessment
- Results of the biocompatibility test
- Assessment concerning the scientific state of the art (technology, medical application/therapy)

This is again followed by a final assessment and conclusion regarding the benefit-risk ratio.

14.9. Initiation of a Clinical Investigation

A combination of clinical investigation and literature search is necessary if no literature data is available or if data from literature does not suffice to prove the safety and performance of the specific medical device.

For higher-risk products (class III and implantable devices), a clinical investigation has to be performed unless it is duly justified to rely on existing clinical data (Directive 93/42/EEC, Annex X, 1.1a and Directive 90/385/EC, Annex 7, 1.2).

Unfortunately the term "clinical investigation" is not defined in detail by the directives or the German MPG (Medical Devices Act). It is therefore recommended to refer to the EU guidance document MEDDEV 2.7/4 or to the standard ISO 14155.

The standard ISO 14155 defines the term "clinical investigation" as "… a systematic investigation in one or more human subjects, undertaken to assess the safety or performance of a medical device".

The objective of a clinical investigation is to gather clinical data that proves safety and/or performance of a medical device. A clinical investigation is always needed if sufficient data cannot be derived from a literature search, or if literature data does not address all aspects required for the full clinical assessment of the specific product.

In some cases clinical investigations are conducted after the medical device has been CE-marked. One example of a clinical investigation being conducted with a CE marked device is an investigation in context of a so-called "post-market clinical follow-up" (PMCF). This follow-up is required according to the EU guidance document MEDDEV 2.12/2 in order to observe the safety and performance of medical devices throughout its lifecycle.

## 14.10. Regulatory Requirements

Requirements related to clinical investigations are described e. g. in the ...

- German Medical Device Act, Medizinproduktegesetz (MPG) §§20, 21, 22, 22a-c, 23, 23 a and b
- German Ordinance on Clinical Trials with Medical Devices, Verordnung über klinische Prüfungen von Medizinprodukten (MPKPV)
- German Ordinance on Medical Devices Vigilance, Verordnung über die Erfassung, Bewertung und Abwehr von Risiken bei Medizinprodukten (Medizinprodukte-Sicherheitsplanverordnung, MPSV) §§ 2, 3 and 5
- German Institute of Medical Documentation and Information, DIMDIV §§ 2, 3, 3a
- Directive 93/42/EEC, Article 15 and Annex VIII and X
- Directive 90/385/EEC, Article 10 and Annex 6 and 7
- Statement of the World Medical Association (Helsinki, June 1964): Recommendations for physicians who are active in biomedical research

Also to be taken into account are the essential requirements of ISO 14155 as well as the recommendations of the MEDDEV guideline 2.7.2., 2.7/3, 2.7/4 and 2.12/2.

---

**Exercise:**

Research requirements related to clinical investigations in MEDDEV 2.7.1., 2.7.2., 2.7/3, 2.7/4 and 2.12/2 and in ISO 14155.

---

14.11. Summary

A clinical assessment is based on the analysis and assessment of specific clinical data of a defined medical device. The objective is to prove the performance and safety of the product during clinical use. This can be done by assessing clinical data from published literature and/or by performing clinical investigations with the respective device.

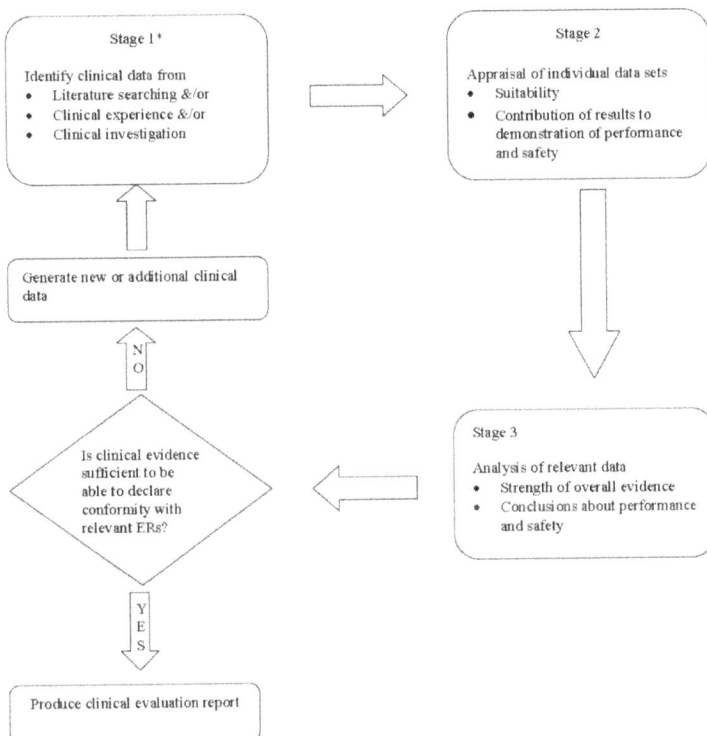

Fig. 14/2: Stages of clinical evaluation (http://ec.europa.eu/health/medical-devices/files/meddev/2_7_1rev_3_en.pdf, page 10)

Two terms must be clearly differentiated: "clinical assessment" and "clinical trial". The clinical assessment in the EU is part of the conformity assessment procedure for medical devices that results in the CE-marking. A clinical trial can be part of a clinical assessment.

14.12. Test Your Knowledge

| | |
|---|---|
| **Q1:** | List the two pathways that can be used for a clinical assessment |
| **A1:** | Clinical investigation(s) and /or literature search |

| | |
|---|---|
| **Q2:** | Does a clinical investigation have to be performed for a non-sterile surgical instrument (class I)? |
| **A2:** | This depends on the availability of clinical data of the product or a similar product. |

| | |
|---|---|
| **Q3:** | A literature search provides only data that proves the performance capacity and safety of a product. Can a clinical assessment be based on these data? |
| **A3:** | No. The literature search should be continued until … |
| | … a certain number of "negative" reports of the product is found |
| | … a small modification of the search criteria does not provide any new reports |

| | |
|---|---|
| **Q4:** | Does a clinical assessment need to be carried out for a non-sterile surgical instrument (class I)? |
| **A4:** | Yes. |

14.13. References

- MDD 93/42/EEC
- Active implantable devices Directive 90/385/EEC
- Directive 2007/47/EC amending Council Directive 90/385/EEC on the approximation of the laws of the Member States relating to active implantable medical devices, Council Directive 93/42/EEC concerning medical devices and Directive 98/8/EC concerning the placing of biocidal products on the market
- MEDDEV 2.7/4 Guidelines on clinical investigation: a guide for manufacturers and notified bodies
- ISO 14155 Clinical investigation of medical devices on humans – good clinical practice
- MEDDEV 2.12/2 Post-market clinical follow-up studies: A guide for manufacturers and notified bodies
- German Medical Devices Act (Medizinproduktegesetz; MPG)

- German Ordinance on Clinical Trials with Medical Devices (Verordnung über klinische Prüfungen von Medizinprodukten MPKPV)
- German Ordinance on collecting, assessing and preventing of MD risks (Verordnung über die Erfassung, Bewertung und Abwehr von Risiken bei Medizinprodukten (Medizinprodukte-Sicherheitsplanverordnung, MPSV)
- German Ordinance on the database-related information systems of medical devices of the Deutsche Institut für Medizinische Dokumentation und Information (DIMDI Verordnung)
- Statement of the World Medical Association (Helsinki 1964), www.bundesaerztekammer.de/ downloads/handbuchwma.pdf
- MEDDEV 2.7.2 Guide for the competent authorities in making an assessment of clinical investigation notification
- MEDDEV 2.7/3 Clinical investigations: serious adverse event reporting under directives 90/385/EEC and 93/42/EEC
- MEDDEV 2.7.1 Clinical evaluation: A guide for manufacturers and notified bodies

## Chapter 15: Risk Management for Medical Devices according to ISO 14971

*Dr. Carsten Rupprath*

15.1. Learning Objective

This chapter gives an insight into the risk management for medical devices and describes some major aspects of risk management.

15.2. Introduction

The risk management is required by the Medical Device Directive 93/42/EEC and is of major importance in the quality management. Risk management is a systematic application of management methods and procedures in order to analyse, evaluate, control and monitor risks.

The international standard ISO 14971 describes the risk management for medical devices and the standard was specifically developed for medical devices. The ISO 14971 standard defines the scope in which the manufacturer can use his experience and expertise, assessment and judgement to systematically assess and limit risks that can occur when using medical devices.

15.3. Risk Management and Medical Devices

The standard ISO 14971 first of all defines the risks for patients but also the risks for user, other persons, other equipment and the environment.

In ISO 14971 a risk concept with two components is described:

1. The probability of the occurrence of harm
2. The consequences of this harm, that is, how severe it might be

The standard describes a process that helps the manufacturer of medical devices to identify hazards associated with medical devices. The process enables the manufacturer to assess, evaluate and control the risks and to monitor the efficiency of this control. The requirements of the standard are valid throughout the entire life cycle of the product. It is not applied in clinical decision processes and it does not define acceptable risk levels. Risk management is a part of the quality management system of a medical device manufacturer. As far as the risk management is concerned, the manufacturer has to establish, document and maintain a system over the whole life cycle of a medical device that identifies, assesses, evaluates, controls and monitors hazards that are linked to the use of a medical device.

The risk management consists of the following parts (EN ISO 14971, 3.1):
1. Risk analysis
2. Risk evaluation (decision of acceptability of risks)
3. Risk control
4. Production and post-production information

The product realisation process that is described in the quality management system (QMS) according to ISO 13485 should be extended by the risk management process.

The top management has to define a policy for determining criteria for risk acceptability (EN ISO 14971, 3.2). These criteria should be based on applicable national and international regulations as well as on relevant international standards and also consider the latest state of the art and known concerns of stakeholders. Moreover, the top management has to support the risk management process by providing adequate resources and qualified staff for this process and to set up a risk management file.

The top management has to check the suitability of the risk management process at planned intervals and to take and document measures if needed. The staff that performs risk management tasks has to have the knowledge and experience they need to carry out the tasks that have been assigned to them. That means knowledge and experience on a defined medical device, the applied technologies or risk management techniques (EN ISO 14971, 3.3). The qualification records have to be kept.

Risk management activities for a medical device have to be planned and documented in a risk management plan that is part of the risk management file. In order to calculate the risks, the probable scope of damage and the probability of the risk should be defined early in the process. An acceptable risk has to be defined beforehand. In case extraordinary decisions need to be taken, e. g. the ALARP (as low as reasonable practical) for a risk is still acceptable, this has to be justified. In the last version of the ISO 14971:2012 the ALARP scope was newly interpreted in "as low as practical". It is recommended to discuss with the Notified Body how they interpret the ALARP scope.

In order to evaluate the efficiency of measures, an acceptance matrix (extent of damage/ probability) can be established before and after the measures. By completing the content of the matrix, the residual risk can be evaluated. Moreover, an addition of the acceptable single risks should be carried out in order to evaluate the overall residual risk. Also, the company has to make sure that data from the product realisation phases, from post-market surveillance as well as data from the application by the customer and service department are incorporated in the risk management.

The risk management plan should contain at least the following elements (EN ISO 14971, 3.4):

- The scope of the planned risk management activities
- Assignment of responsibilities and authorities
- Requirements for review of risk management activities
- Criteria for risk acceptability (based on the manufacturer's policy for determining acceptable risks)
- Verification activities
- Activities related to collection and review of relevant production and post-production information

Changes in the risk management plan have to be documented. The manufacturer has to establish and maintain a risk management file for every medical device. The risk management file has to enable the traceability of each identified hazard in the risk analysis, the risk evaluation, the implementation and the verification of risk control measures as well as the assessment of acceptability of residual risks (EN ISO 14971, 3.5).

The risk analysis has to be established for each medical device. The implementation of planned risk management activities and the results of the risk analysis have to be documented in the risk management file.

The risk analysis has to consist of the following elements (EN ISO 14971, 4.1):

- Description and identification of the analysed medical device
- Identification of the persons and organizations who carried out the risk analysis
- Scope and date of the risk analysis

The manufacturer has to document the intended use of a medical device as well as foreseeable abnormal use. In a next step the manufacturer has to evaluate the qualitative and quantitative characteristics which are defining the safety of a medical device – as well as their limits (EN ISO 14971, 4.2).

The manufacturer has to identify and document the known and foreseeable risks of its medical device and has to add this information to the risk management file (EN ISO 14971, 4.3). As far as the identified hazards are concerned, the risks have to be evaluated and documented taken into account all available information. In case of hazardous situations where the risk cannot be clearly identified, possible consequences have to be listed in the risk evaluation and risk control. All applied systems of the qualitative as well as quantitative classifications of probabilities of harm or the severity of harm have to be recorded as part of the risk management file (EN ISO 14971, 4.4).

For each identified hazardous situation, the manufacturer shall decide based on the criteria of the risk management plan, if a risk reduction is required. In case risk reduction is not required, the risk control process is complete. The results of the risk evaluation shall be documented in the risk management file (EN ISO 14971, 5).

Risk control activities will be performed, if a risk reduction is required (EN ISO 14971, 6.1). The manufacturer has to identify risk control measures that reduce risk to an acceptable level via the risk control option analysis (EN ISO 14971, 6.2).

The manufacturer has to conduct one of the following risk control options in the stated order:

1. Inherent safety by design
2. Protective measures in the medical device itself or in the production process
3. Information for safety

The chosen risk control measures have to be documented in the risk management file. In case the risk control option analysis clearly shows that a risk reduction is not practical, the manufacturer has to carry out a risk-benefit analysis to evaluate the residual risk.

The implementation of risk control measures has to be verified and recorded in the risk management file. Moreover, the efficiency of the risk control measures has to be verified (EN ISO 14971, 6.3). After implementation of the risk control measures, the residual risks have to be evaluated applying the criteria of the risk management plan. In case that the residual risks are evaluated as not being acceptable, further risk control measures have to be applied. The manufacturer has to decide what residual risks that are judged acceptable, he is going to list in the instructions for use thus informing his customers and users (EN ISO 14971, 6.4).

In case that the residual risk is not acceptable and a further risk control is not practical, the manufacturer has to check the data and literature sources in order to evaluate whether the benefits of the intended use outweigh the residual risk (EN ISO 14971, 6.5). If the data do not support this conclusion, the risk of applying the medical device remains inacceptable and the product cannot be distributed and will not be certified by the Notified Body. In case the benefits outweigh the residual risks, the overall residual risk is acceptable and the product might be distributed (for higher medical decice classes after CE certification by a Notified Body).

The influence of the risk control measures has to be checked as well with a special focus on the question if they generate new hazards or hazardous situations or have an influence on already identified hazardous situations (EN ISO 14971, 6.6). All new or increased risks have to be evaluated and – if possible – reduced/avoided.

Finally, the completeness of the risk control has to be documented by the manufacturer. He has to state that all identified, hazardous situations have been taken into consideration (EN ISO 14971, 6.7). After implementing and verifying all risk control measures, the manufacturer has to decide whether the overall residual risk of the medical device is acceptable or not according to the defined criteria of the risk management plan (EN ISO 14971, 7). The overall residual risk is evaluated.

Further data may be taken into consideration to decide whether or not the medical benefits of the intended use outweigh the overall residual risk. The overall residual risk can be judged acceptable if the conclusion can be drawn that the medical benefits outweigh the overall residual risk. If that is not the case, the residual risk remains inacceptable and thus the product cannot be distributed. All results and decisions of the risk management process have to be documented (EN ISO 14971, 8).

Before distributing the medical device, the manufacturer has to conduct an evaluation of the risk management process. This evaluation shall guarantee that a risk management plan has been implemented, that the overall residual risk is acceptable and that adequate methods are applied in order to collect the relevant production and post-production information. The results of this evaluation are to be documented in the risk management report that is part of the risk management file. The evaluation of the risk management process has to be carried out by qualified and authorized persons.

Moreover, the manufacturer has to establish and maintain a system that monitors the device during the production and post-production phases (EN ISO 14971, 9). The manufacturer has to take into consideration that information of e. g. the operator, user, service technician, sales representative etc. is collected and processed and that compliance also with new and revised standards is ensured. The system should also collect publicly available information of similar medical devices. This information has to be evaluated regarding their relevance to safety especially as far as unidentified hazards are concerned or whether the overall residual risk of the product is acceptable.

The influence of new data/information on already implemented risk management activities has to be evaluated and used as input to the risk management process. In case there is a potential that the residual risk acceptability has changed, the impact on previously implemented risk control measures shall be evaluated (EN ISO 14971, 9).

## 15.4. Summary

It can be stated that ISO 14971 defines a basic structure that can be used to evaluate and control the risks when applying a medical device. The defined process of ISO 14971 supports the manufacturer in identifying hazards that are linked to the use of the medical device. These risks can be evaluated and controlled and the effectiveness of the controls can be monitored.

## 15.5. Test Your Knowledge

| | |
|---|---|
| **Q1:** | Why is the identification of the persons – involved in the risk analysis – of importance? |
| **A1:** | A risk evaluation can only be conducted by competent and trained staff. In internal and external audits, this important aspect is often evaluated by identifying the involved persons as well as their CV and training curriculum. |

| | |
|---|---|
| **Q2:** | That a product has to be certified by a Notified body before being placed on the market depends on ...? |
| **A2:** | ... the classification of a medical device. Class I medical devices do not need a certification whereas medical devices of e.g. class IIa, IIb and III need to be certified by a Notified Body. |

| | |
|---|---|
| **Q3:** | Why is it important to take into account new data/information in the risk management of an established medical device? |
| **A3:** | Further risk, e. g. long-term risks can always occur. New information (for example from customer feedback or of a similar product of a competitor) can influence the risk-benefit assessment significantly. |

## 15.6. References

- EN ISO 14971: Medical devices – Application of risk management to medical devices
- Risk management of medical devices – Guideline for applying DIN EN ISO 14971:2001, BVMed

- Medical devices: Risk management in the life cycle of a product according to ISO 14971:2007: Vigilance system by Jürgen P. Bläsing, TQU Verlag
- Requirements to medical devices: practical guideline for manufacturers and suppliers by Johann Harer, Carl Hanser Verlag GmbH & CO. KG

# Chapter 16: Regulatory Risk Management

*Dr. Stefan Menzl, Dr. Carsten Rupprath*

## 16.1. Learning Objective

This chapter focuses on the main aspects of the regulatory risk management. The control of the regulatory risks is important for an unrestricted, quick and sustained market access.

## 16.2. Introduction

Risk management does not only refer to products but also to the regulatory landscape where risks arise that should be avoided.

## 16.3. "Cooperation & Communication" – Which Partner to Address?

In order to minimize risks from a regulatory view, a close cooperation of all internal and external partners is feasible and highly purposeful. In this chapter, the most important cooperation partners for regulatory risk management will be presented.

### 16.3.1. Notified Body

The right choice of the Notified Body is of great importance for a manufacturer of medical devices (compare chapter 11). The main criterion for choosing a Notified Body certainly is not only the costs for CE-markings, change notifications and quality certifications of the Notified Body. Of great importance for a manufacturer are other aspects of a Notified Body, f. e. the knowledge of the products or if the Notified Body's regulatory department will represent the manufacturer's interests. As a lot of Notified Bodies nowadays face a high workload, it is essential for the manufacturer to be able to quickly reach a Notified Body employee via phone or email and that the manufacturer's questions are promptly answered and tasks completed in due time. A Notified Body can also support a manufacturer in special cases, f. e. when an extraordinary enquiry from Competent Authorities of the Middle East or Asia has to be answered that means that a letter or statement of a Notified Body is requested. Such a service is not rendered by all Notified Bodies – or at least not in due time even though this would be highly appreciated by the manufacturer.

A manufacturer should choose more than one Notified Body for the certification of his products. This of course depends on the size of the company. For small companies it is feasible to work with only one Notified Body. To work with several Notified Bodies has the advantage that one can compare the results and services of the Notified Bodies. If there is a great bias or a significant decline in services, it is quite easy to exchange the Notified Body. If a manufacturer works together with several Notified Bodies, he personally knows his contact partners and exchanging one Notified Body for a proven reliable other one is an easy decision to take. Therefore, big manufacturers should definitely work with several Notified Bodies. The requirements are the same, but some of the above mentioned criteria can change quickly within the Notified Body. Having more than four to five Notified Bodies is practically of no advantage. Moreover, the risk of dissipation is high and the handling of such a high number of Notified Bodies and contact persons is difficult. The average number of two to three Notified Bodies for big manufacturers with many products has proved successful (The list of Notified Bodies can be found on the following website: http://ec.europa.eu/enterprise/newapproach/nando/index.cfm?fuseaction=notifiedbod y.main)

### 16.3.2. Cross-functional Cooperation at the Manufacturer

Regulatory risks can be minimized by an enhanced cross-functional cooperation with different departments within a company. Regulatory Affairs can be involved in the choice of suppliers for example in order to verify whether or not necessary certificates exist. Regulatory Affairs can also help avoiding regulatory risks when choosing a raw material supplier: if the use of material of animal origin can be excluded and a class III classification of the product thus avoided this will minimize the risk and expenditure for the company. Regulatory Affairs can also support the planning process for clinical studies in order to meet the requirements of the Notified Body. When planning to register a new product or submitting a change notification, the early involvement of Regulatory Affairs can be of great advantage to the manufacturer. Regulatory Affairs can develop the right registration strategy for a given case. Also the strategy planning of Marketing and Sales can profit from regulatory input. The question whether or not it is reasonable to register a product in a defined country can be answered by estimating of the registration costs and follow-up costs. Therefore, f. e. a requested clinical study might not make sense when only a small profit and small turnover can be expected in the medium term in a specific country.

As far as registration processes are concerned, Regulatory Affairs can prevent "surprises" by f. e. meeting with the Notified Body before submitting the new product. When presenting the strategy and plans for the new product to the Notified Body, one will get feedback from the Notified Body that will result in being able to draw conclusions. For example, in case a new product is similar to an already existing product with CE-mark, it can be found out whether a separate clinical study has to be carried out or if an existing clinical study for the already registered product – from the point of view of the Notified Body – is sufficient.

Backup strategies of the Regulatory Affairs department of a manufacturer can help in advance to prevent frequently occurring problems. Raw materials from third-party suppliers bear a certain risk for a manufacturer because the supplier might take a raw material that is used by the manufacturer off the market due to various reasons. It is important for the manufacturer to develop strategies for this situation. An anticipating Regulatory Affairs department can enlist additional suppliers or raw material manufacturers, thus helping the manufacturer to tackle unforeseeable events of such kind that would have a regulatory impact and result in f. e. time-consuming change notifications.

### 16.3.3. Distributors
Manufacturers who sell their products via distributors should be aware of the advantages in case they are also the holder of the registration certificate. If the distributor owns the registration, the manufacturer depends on this distributor. If – in this case – the manufacturer decides to change this distributor, he has to apply for a new registration and until this is granted, he cannot sell his products in this market. By being the holder of the national registration, the manufacturer can prevent this unfortunate case, because he can exchange the distributor quite easily without interrupting the access of his product to the market. This manufacturer registration can be obtained via the regulatory department of the manufacturer or via a freelance consultant.

### 16.4. Summary
It is of great importance to the manufacturer to be able to plan as precise as possible at which point in time a registration for a new product is granted (in order to optimize

the market access). Regulatory Affairs can facilitate this market access by providing various, reality-based scenarios regarding the registration (best case, medium case or worst case).

The development of these registration scenarios is made possible through an intensive exchange with the Notified Body. This enables the department to plan as precise as possible the point in time when the registration is granted and the earliest date possible to have the product placed on the market. By developing realistic registration scenarios the risk of a "delayed registration" as well as a "premature registration" can be minimized. Both scenarios would be damaging for the manufacturer for different reasons: a) a "delayed registration" leads to a decline in planned turnover or image of the manufacturer as a competitor could be first on the market with a similar product and b) a "premature registration" would mean that the product is not ready to be sold as the market access is not prepared and the product not yet available.

Moreover, the target countries and target regions for a new product should be planned together with Regulatory Affairs in order to meet all national requirements (see chapter 18) and languages in due time without jeopardizing the planned launch of the product.

The regulatory department of an international company ideally has to structure the documents in a way that it can be used worldwide. This strategy saves time and significantly shortens the time to market entry. Furthermore, the regulatory department has to take into account standards and tests that go beyond the ISO standards. This anticipating planning enables a significantly faster market entry in countries that developed their own standards and expect these national standards to be met. Especially the specific test requirements for China should be taken into account during product development.

It is the task of the regulatory department to regularly monitor the registration requirements off all countries in which the manufacturer sells his products. This ensures an uninterrupted market access of the already registered products and a fast market entry for new medical devices. An ongoing review and monitoring of the requirements is key to a sustained market presence of a manufacturer.

16.5. Test Your Knowledge

| | |
|---|---|
| **Q1:** | Why is the cross-functional cooperation key for a sustained market access of a company? |
| **A1:** | In order to enter the market as soon as possible, involving the regulatory department at an early stage is important because obstacles regarding the registration can be avoided and an ideal registration strategy can be developed. Risks that might jeopardize a sustained market presence can be anticipated and backup strategies developed. |

| | |
|---|---|
| **Q2:** | Think of reasons why a "premature registration" might be of disadvantage to the manufacturer? |
| **A2:** | Getting a registration for a new medical device is always good news for a company. But a "premature registration" can also have negative aspects. The planned number of the new product may not have been produced yet, the warehouse does not have enough stock to supply the customers with the product, Marketing has not started yet to promote the product and thus the manufacturer does not benefit from the premature registration. If the point of time of the registration could have been better predicted, the product could have been sold already in higher volumes. Therefore, a "premature registration" most of the time means missing a chance because no manufacturer is able to arrange an significantly earlier date for the market entry. If a company often has "premature registrations" the processes of predicting and assessing the registration approval has to be reviewed. An, as precise as possible, prediction of the registration enables the manufacturer to ideally plan his resources and not to miss a market opportunity. |

16.6. References

- List of the Notified Bodies:

  http://ec.europa/enterprise/newapproach/nando/index.cmf?fuseaction=notified body. main

# Chapter 17: Interactions with other Departments – Inside or outside the Company

*Dr. Sibylle Scholtz, Dr. Stefan Menzl*

## 17.1. Learning Objective

In this chapter you will learn to understand the complex cooperation of Regulatory Affairs (RA) with other departments inside and outside the company.

## 17.2. Introduction

There are different ways a Regulatory Affairs department is positioned in a medical technology company that results in different philosophies and objectives. Especially in the interface between different departments, harmonized work processes are important in order to avoid frictional losses that could lead to delays in modifying a medical device or even to noncompliance. This could result in higher costs and a higher risk for a company.

## 17.3. Role of Regulatory Affairs

In most medical technology companies Regulatory Affairs ensures that quality, safety and effectiveness of a medical device complies with the legal requirements of the target markets. That makes RA an integral part of such a company.

Internally, RA is involved in design & development, production, clinical assessment, promotion as well as post-market activities. This means that the whole life cycle of a product is covered.

Externally, RA often is the contact partner between company, Notified Bodies and competent authorities.

It definitely makes sense to involve RA as early as possible in the design & development of a medical device in order to put together relevant dossiers that are submitted to the competent authorities for assessment. RA is also involved beyond the design & development phase and e. g. has to take care of the complaint management and further life-cycle development.

The RA employees play a key part when defining the product design strategy, especially in highly controlled international markets that are getting more and more complex. They also play an important operative role when mediating reliable processes or interactions with authorities.

Registration processes are based on scientific principles and therefore are very dynamic. As the technical development and standards are constantly developing and changing, RA faces challenges and therefore RA offers a wide range of career perspectives.

In a world that is getting more and more complex, the exchange with the approval authority is a key strategic objective for the company because the authority can give valuable hints about feasible strategies for the product design phase, authorise clinical studies and enable market entry via an approval of the product.

## 17.4. Definition of the Scope of RA Activities

Usually, RA employees collect information on approval requirements and make these available to research and design departments. These requirements change constantly, and therefore have to be monitored closely by Regulatory Intelligence as they are key to the success of a company that wants to sell its products in different markets (also compare chapter 10).

Moreover, RA also collects information on international, regional or local requirements. International standards are important in order to ensure regulatory compliance with the requirements in key target countries worldwide. Compliance e. g. with a deviating standard in China can be the prerequisite to be allowed to sell a product in the huge Chinese market.

Local standards often deviate only from the applicable version of the current international standard. Sometimes they differ in applicable tests, specifications or criteria.

On the basis of general approval requirements and applicable standards, RA employees can define an approval strategy for a company and a defined product.

The definition consists of

- Defining the target countries
- Defining the sequence of the needed approvals (as referring in one country to approvals in other countries can help getting the approval with low effort)

The choice of material can have significant influence on the review work and complexity of an approval submission. For example, if a manufacturer chooses a coating of a starting material of animal origin (e. g. pig heparin), the coated medical device will be classified to class III. This results in more complex requirements to prove the conformity with the essential requirements according to MDD 93/42/EEC and it will take longer to get the CE-mark. The choice of alternative materials could lead to a lower classification and thus to a faster market entry.

The choice of suppliers can also influence the time to market entry and the effort to get approval. If a supplier does not have a certified quality system, this fact could lead to significant challenges when trying to get the approval for a product.

The choice of a design and manufacturing site has to be well considered. If the manufacturer has a production site in a country in which an approval is not planned or will not be granted at an early point in time, this could negatively affect the approval in some markets in Asia and Latin America. In some countries of these regions, so-called "free sale certificates" (FSCs) or "certificates-to-foreign-government" (CFGs) are the prerequisite to get an approval. The competent authority of the country, in which the manufacturer resides, confirms via these certificates that a product is allowed to be placed on the market. The absence of this certificate means that an approval cannot be achieved in some countries.
Some countries require that these certificates have to be issued by the country in which the product is manufactured. Other countries accept when these certificates are issued by the country of the legal manufacturer. The sequence of the needed approvals therefore has to be planned with great care.

On the basis of an approval in the US and in Europe (CE-mark) it is much easier to get registrations in a lot of countries worldwide. The competent authorities of these countries often rely to a certain extent on the assessments that FDA and Notified Body have already conducted.
Moreover, Regulatory Affairs also ensures that all other departments of the company know what content of the technical documentation is required and provide this information in due time. The active cooperation when putting the technical

documentation together (maybe also in form of a global approval dossier) often is the core business of Regulatory Affairs.

Another important task is choosing a Notified Body – this decision is taken by several departments: Regulatory Affairs, Quality Assurance, Clinical Research and Development. As the same Notified Body is responsible for product certification as well as for the certification of quality systems, this decision of choosing a Notified Body often is for a long-term period of time.

When choosing a Notified Body, not only the accreditation for a defined product category is of high importance but also the knowledge and experience with the relevant technology, the easy accessibility of the NB staff and, if appropriate, further services that the NB can offer in countries outside the EU (see chapter 11).

Regulatory Affairs staff members often work closely together with the departments Research & Development as well as Clinical Investigation and Quality Assurance in order to choose appropriate test methods in order to be able to prove compliance with a standard (for the risk assessment).

It is important to specify with these departments the sequence or prioritisation of single parts of the technical documentation depending on a modular approach of the submission. For example, when long-term stability studies are required, it makes sense to submit those parts of the TD to the authorities that are already available. When the stability data become available, this module can be submitted then. The assessing authority or the NB can assess these data exclusively – that leads to a shorter time required for the assessment.

Another strategy to reduce the assessment time of a competent authority or a NB and to avoid time-consuming, unexpected surprises (= demands of any kind by the authority) is to involve the authority and the NB already at an early stage.

In so-called pre-submission meetings one can address all issues, e. g. whether the planned strategy by the company is approved by the authority/NB and what additional requirements they may lay down. These meetings also need close cooperation of all involved departments of a company.

One of the core tasks of RA also is the assessment of the completeness of the design dossier. If there is any information missing, the departments that are responsible to provide this information have to be contacted to fill in the missing information. Part of the assessment of the design dossier is also assessing whether or not high level documents (e. g. reports) are coherent with the underlying raw data.

The product life cycle of a medical device is relatively short compared to that of a medicinal product. Products and technologies develop very quickly and this results in regular modifications of the design, process flow, materials, sterilization and manufacturing. All these changes have to be assessed regarding the registration. The objective of this assessment is to monitor how these changes are documented whether or not the approval authority or NB has to be informed and whether they have to consent to these changes before these changes are implemented.

Also when defining the product labelling and the product information, RA and other departments work closely together. Their task is to ensure that the info on the product labelling complies with the approved information. This is of high importance for performance characteristics, indication and warnings. Usually it is RA that is in close contact with the NB or approval authority at submission and if later on questions by the authority have to be answered. This is to ensure that there are no discrepancies in the communication with authorities.

A lot of products are marketed in different countries in different variants. Again, it is up to RA to ensure – in close cooperation with Sales and Marketing – the management of local product variants and that only these variants are released and promoted that are approved for the respective country. The same is true for differences in the specification of products in markets or differences in materials or suppliers.

Like other functions, e. g. Quality Assurance, Sales, Marketing and Finance, RA is also actively involved in the release of products to defined markets. The objective is to only release those products that comply with all requirements of that country, for example language and labelling requirements.

When a product is released and promoted in a country, the responsibility of RA does not cease. In this phase, RA gives input to a post-market surveillance plan and to a post-market clinical follow-up plan (as required by MDD 93/42/EEC). Moreover, RA is involved in the assessment of complaints, adverse events and trends. If adverse events meet the notification criteria, in some companies it is RA that is going to notify the authorities and the NB. Together with Quality Assurance, RA is in contact with the authorities and the NB in case of field safety corrective actions or product recalls. Before an information or warning is passed on to product users the content of the information/warning is assessed by RA.

Numerous change notifications and resulting delays until the product is approved can be avoided by early planning or by describing planned modifications already in the original submission (e. g. as a variant). This way, substantial changes can be avoided. Adequate strategies are developed in cooperation with R&D, Sales and Marketing.

Another focus is the assessment of the product advertising and the compliance with the approved indication. In this area, RA works closely together with Marketing by providing their scientific competence and knowledge on country-specific, approved performance criteria. Moreover, RA is involved in defining the audit strategy for the company and critical suppliers. This takes place in close cooperation with Quality Assurance and also comprises the identification of critical areas for an audit. Feedback from product complaint management (feedback from the market regarding problems that occurred with a product) can provide valuable hints.

## 17.5. Summary

RA plays a central role in the design phase of a medical device as well as after the market release. Of crucial importance for tackling the wide range of tasks is a close cooperation on a basis of mutual trust with nearly all departments of a company.

## 17.6. Test Your Knowledge

| | |
|---|---|
| **Q1:** | Name 3 departments inside a company with which RA works closely together. |
| **A1:** | Quality Assurance, Research & Development, Sales, Marketing, etc. |

| | |
|---|---|
| **Q2:** | With what external central institution does RA work closely together? |
| **A2:** | With a Notified Body. |

| | |
|---|---|
| **Q3:** | Does RA support Quality Assurance with product recalls? |
| **A3:** | Yes. |

| | |
|---|---|
| **Q4:** | How can the choice of manufacturing site influence the registration strategy? |
| **A4:** | If a company chooses a manufacturing site in a country where a registration can be acquired at an early point in time, a free sale certificate can be issued by this country. This facilitates the registration in many other countries of the world, especially in Asia and Latin America. |

## 17.7. References

- MDD 93/42/EEC of June 14, 1993
- Directive 90/385/EEC of June 20, 1990 on the approximation of the laws of the member states relating to active implantable medical devices
- Directive 98/79/EC of October 27, 1998 on IVDs

Interactions with other Departments

**Chapter 18: Selected National Peculiarities – Notification Obligation for Medical Devices that are Placed on the Market**

*Dr. Carsten Rupprath*

18.1. Learning Objective

This chapter focuses on information on national peculiarities that have to be considered when distributing medical devices into the EU.

18.2. Introduction

The CE-marking of a medical device by the Notified Body was meant to enable the free distribution of a product in all EU member states. Some EU countries have established additional registration obstacles because they were not satisfied with the requirements of the CE-marking and also wanted to have more control over their national markets.

The delayed implementation of the European medical device database EUDAMED also encouraged the implementation of additional national registration databases because national Competent Authorities had the impression that they did not have adequate information which medical devices were distributed by which manufacturers in their country. They were convinced that they were not able to properly fulfil their responsibility to protect their citizens from potentially unsafe products without additional requirements.

This impression cannot be shared by medical device manufacturers because all medical devices that are distributed in Europe have to undergo the CE-marking process. Due to the additional registration the national Competent Authorities have more information available on products that are distributed in their country but this does not result in a higher safety for the population. Moreover, one has to state that these additional national registration systems are a lucrative source of income for the national Competent Authorities.

## 18.3. Which Countries are Affected

### 18.3.1. EU Countries

In many countries (e. g. Denmark, Finland, Ireland, Sweden and Hungary) the national Competent Authorities have to be informed about the distribution of medical devices class I, procedure packs and custom-made devices by the manufacturer or his Authorized Representative having their headquarters in that country. The national Competent Authorities want to have an overview about products, manufacturers and Authorized Representatives in their area of responsibility.

---

**Exercise:**

Go to the internet and look for peculiarities of the countries that are described in the following sections.

---

### 18.3.2. Italy

Since 2007, there is in Italy a national registration database of the Italian health ministry ("repertorio"). In this database, all CE-marked medical devices (of class I to III) have to be entered before the products may be distributed. The fee of 100 € that has to be paid initially as well as the link to the Italian reimbursement system has been cancelled. For the registration, the CE-certificate, the Italian packaging/labeling and the instructions for use of each medical device have to be uploaded in the database. Moreover, further information (e. g. sterilization methods or GMDN codes) is needed to fill in the online forms. Clinical data is not needed anymore for the database registration. The additional time needed for the database registration should be taken into account when a product launch is planned in Italy. Custom-made medical devices or IVDs do not have to be registered.

### 18.3.3. Ireland

A manufacturer or his Authorized Representative has to register at the Competent Authorities of the EU country in which he is residing, before initially placing a medical device on the market. Moreover, medical devices of class I have to be registered at the Competent Authorities of an EU member state by a manufacturer or his Authorized Representative. Also, a fee has to be paid. A reason for this requirement that has been implemented by all member states, is the control of class I products that only need a Declaration of Conformity by the manufacturer (and no marking by a Notified Body) for being distributed in the EU. In **Ireland**, the Irish Medicines Board (IMB) has established an online database for class I products, custom packs and IVDs, in which domiciliary companies or Authorized Representatives have to register.

### 18.3.4. Belgium

In Belgium, there is a special tax on medical devices that means that distributors have to pay 0.05 % of the turnover per year to the Ministry of Health.

### 18.3.5. France

In France, a national registration of professional, reimbursable medical devices in a database became necessary in 2013 with the implementation of the Xavier Bertrand Act. Private and public hospitals are only allowed to buy medical devices that are registered in the national database. For this registration, technical documents (specifications etc.) and especially clinical documents (e. g. clinical trials) are needed. The registration in the database is only valid for a limited period of time but can be extended. The French health authority (Haute Autorité de Santé) is planning to publish a list of all affected products (probably higher-class, higher risk medical devices, e. g. invasive and implantable MDs). This registration is probably linked to a fee per product registered (information of November 2013, size of fee not known yet). Moreover, there is in France – like in Belgium – an additional tax (0.25 % of the turnover) for medical devices and IVDs that has to be paid by the manufacturer or the distributor if the overall turnover in France is above 763.000 €.

### 18.3.6. Germany/Austria

In Germany and Austria, a medical device manufacturer or his Authorized Representative have to appoint a local security officer in accordance to § 30 (Germany) and § 78 (Austria) of the Medical Devices Act (MPG). The local security officer has to collect reported risks of medical devices, to assess them and to coordinate all necessary measures. He is responsible for the fulfilment of the obligation to notify if these reports are about medical device risks (see chapter 20). Also the function of medical device consultant has only been legally fixed in Germany (§ 31 MPG) and Austria (§ 79 MPG). The medical device consultant has to document information of health care professionals on side effects, interactions, malfunctions, technical defects, contraindications, tampering or other risks. He then has to immediately submit his written comments to the responsible person or the security officer for medical devices (see chapter 20).

In Germany, there is a general notification obligation according to § 25 MPG for the responsible entity (§ 5: 1 + 2 MPG) if it is located in Germany and if medical devices are placed on the market for the first time (except medical devices of § 3: 8 MPG). The notification about medical devices and IVDs together with the initial placing on the market (§ 25) and the security officer (§ 30 MPG) have to be placed according to the DIMDIV – the regulation of the database-supported information system of the Deutsche Institut für Medizinische Dokumentation und Information (German Institute for Medical Documentation and Information) – via an internet-based collection system (www.dimdi.de/static/de/-mpg/ismp/anzeige/index.htm).
The registration of medical devices in this information system is free of charge.

Every manufacturer residing in Austria has to register in the Austrian Medical Device Register before starting distribution. This registration is mandatory according to § 67 MPG for all natural and legal entities, commercial law partnerships or registered acquisition enterprises (residing in Austria) that are responsible for the initial placing of medical devices on the market of the European Economic Area (EEA). Manufacturers of medical devices (also manufacturers of custom-made devices) or IVDs from countries outside the EEA and the Authorized Representatives of medical devices and IVDs are subject to registrations as well as their products that are placed on the market for the very first time. When registering, also the security officer has to be appointed (§ 78 MPG). The Austrian Register for Medical Devices was established by the ÖBIG (Austrian Health Institute) on January 2, 2002 by order of the Ministry for Health, Family and Youth (MBMGFJ) and can be found under the following link http://medizinprodukte.oebig.at.
The obligatory registration of medical devices and IVDs is solely possible via this link and has to be performed by all required companies/persons (the registration is free of charge).

## 18.3.7. Portugal

In Portugal, there is also an obligation to register all medical devices of class IIa, IIb and III that are distributed in Portugal in a database. For registration, the CE-certificate, the Declaration of Conformity, packaging/labeling and instructions for use (IFU) are required.

## 18.4. Summary

Manufacturer of medical devices should check with the national Competent Authorities of an EU member state if there are special regulations or additional requirements before distributing the products in this country. The national requirements of the member states change regularly (Cave: the information in this chapter is not exhaustive!).

Generally, it can be stated that for several years there is a trend in Europe to additional national regulations that make the free distribution of medical devices more difficult inside the EU even after the CE-marking by a Notified Body. This is irritating because there is no additional safety or protection regarding potential risks for the patient/consumer. Moreover, some states use the fees as a lucrative source of income. Especially the national registration procedures in Italy or the relatively new ones in France and Portugal result in an unnecessary work load for medical device manufacturers. If this trend prevails, the free distribution of products within the EU will no longer be possible or lead to delays for small and mid-sized companies and this cannot be in line with the interests of European citizens and patients.

## 18.5. Test Your Knowledge

**Q1:** What are custom-made devices?

**A1:** Custom-made devices are tailor-made medical devices that are manufactured – due to requirements of a physician or a health care specialist – individually for a customer. These are for example dental prostheses or hearing aids. Custom-made devices are not subject to requirements of conformity of the CE-marking and therefore often have to be registered with the Competent Authorities.

| **Q2:** | What are procedure packs? |
|---|---|
| **A2:** | Procedure packs contain medical devices that are needed for example for a specific eye surgery. These products often are single-use products that are supplied without a primary packaging in an overall combined packaging. Procedure packs facilitate the supply inventory of hospitals because all medical devices are supplied in one container. Moreover, the preparation time for the users is shorter. |

## 18.6. References

- "Repertorio", Italy; www.salute.gov.it/dispositivi/paginainternasf.jsp?id=499
- IMB (Irish Medicines Board); www.imb.ie
- HAS (Haute Autorité de Santé); www.has-sante.fr
- MPG (Germany): Medical Devices Act
- www.gesetze-im-internet.de/bundesrecht/mpg/gesamt.pdf
- MPG (Austria): Medical Devices Act; www.bmgfj.gv.at/cms/home/attachments/7/9/0/CH1095/CMS1228212263945/medizinproduktegesetz_kompiliert.pdf
- DIMDIV (Regulation on database-supported information system on medical devices of the German Institute for Medical Documentation and Information); www.gesetze-im-internet.de/dimdiv/index.html
- DIMDI (Deutsches Institut für Medizinische Dokumentation und Information); www.dimdi.de
- ÖBIG (Österreichisches Bundesinstitut für Gesundheitswesen); http://medizinprodukte.oebig.at

## Chapter 19: Promotion of medical devices

*Dr. Carsten Rupprath, Myriam Becker*

### 19.1. Learning Objective

This chapter gives an overview of the national requirements that have to be considered in the promotion of medical devices.

### 19.2. Introduction

There is no European regulation for the promotion of medical devices that has to be applied in all member states and promotion is not covered by the MDD 93/42/EEC. The promotion of medical devices is regulated by every member state. Especially for Southern Europe there are some requirements to consider.

### 19.3. National Regulations

### 19.3.1. Italy

In Italy, the promotion of professional medical devices to patients is not allowed. Companies have to ensure that advertising activities are addressed to health care professionals. The promotion to health care professionals has to be approved by the Italian health authority. The fee to be paid is 300 € per advertising text, per product and per channel.

### 19.3.2. Spain

There are similar regulations in Spain like in Italy. Like in Italy, the promotion of professional, reimbursed medical devices to patients is not allowed. The promotion to health care professionals has to be authorized by the regional, competent health authority. The promotion to health care professionals has to include all information on the conformity of the medical device to the requirements as well as information on contraindication and side effects.

When promoting non-professional medical devices to consumers or patients, reference to the health authority and recommendations from scientists or health care professionals cannot be used.

### 19.3.3. France

The new Xavier Bertrand Act in France also limits the promotion of reimbursed professional products to patients. Class I and IIa products are excluded from this law and are allowed to be promoted to patients (www.legifrance.gouv.fr/affichTexte.do?-cidTexte=JORFTEXT000026835146&fastPos=47&fastReqId=1710464764&categorieLien=id&oldAction=rechTexte).

Moreover, a list of products was published that have to be authorized by the French health authority when promoted and placed on the market (www.legifrance.gouv.fr/affichTexte.do?cidTexte=JORFTEXT000026451423&fastPos=16&fastReqId=1473494667&categorieLien=id&oldAction=rechTexte).

The following material has to be submitted to the authorities for approval: the promotional material, an approval form, reference material and instructions for use. For every submission the ANSM is said to charge a fee of 500 € per promotional material.

### 19.3.4. Germany

In Germany medicinal products as well as medical devices are subject to the Heilmittelwerbegesetz (HWG, the German Law on Advertising of Medicinal Products). But the requirements for medical devices are not as strict as the ones for medicinal products. For example for medical devices the promotion containing recommendations from scientists or VIPs is allowed. Moreover, patient cases, people wearing work clothing, pictures of a disease, mode of action as well as before and after comparisons are also allowed (Quelle BVMed: Manfred Beeres, Angela Kotter, Kommunikation und Werbung für Medizinprodukte: Was ist erlaubt, was ist verboten? www.bvmed.de/themen/kommunikation/Werbung_fuer_Medizinprodukte/article/Praxisleitfaden_HWG.html).

For medicinal products the following is not allowed according to HWG §3 and § 11:

- Misleading promotion
- Manipulative promotional messages
- Statements that provoke fear in patients
- Recommendations or appraisals from third parties
- Promotion addressed to minors (below 14 years of age)
- Concealing or playing down of application risks

The general press law defines that false assertions are not allowed to be published. The publication of an opinion, however, does not have to be proven. A personal opinion can be "wrong". The prohibition of misleading advertising by the competition law has to be considered that requires that promotion should be honest and true. The slander of a competitor by spreading of untruths is prohibited.

The 16th amendment of the German Arzneimittelgesetz (AMG, Medicines Act) also saw further liberalisations in autumn 2012 regarding the HWG for medical devices. Now, the promotion of patient testimonials – that are objective and not misleading – is allowed.

### 19.3.5. Austria

According to §104 (layman promotion) of the Austrian Medizinproduktegestz (MPG, Medical devices Act) promotion of medical devices is NOT allowed that

1. Are subject to prescription according to §100 (medical devices that are applied by layman users and that may lead to hazards – directly or indirectly – to the health of human beings even when used properly or that are often or to a significant extent not used properly and consequently endangering health)
2. Are subject to be used by health care professionals on patients or
3. Can only be used by consumers – according to the instructions for use – when supervised by a healthcare professional.

According to § 107 of the Austrian MPG promotion for medical devices that is addressed to consumers has to contain

- The labelling and the intended use of the MD
- An information of the rational use of the MD
- An easily noticeable hint in case the MD may provoke side effects or in case special safety precautions have to be taken.

In section 2 of § 107 it is required that promotion of a medical device to consumers has to contain a clear instruction to carefully read the instructions for use and to consult a health care professional before applying the medical device. If acoustical or audio-visual promotional channels are being used this hint needs to be clearly perceivable.

### 19.3.6. Switzerland

In Switzerland the consumer promotion of medical devices that have to be prescribed or applied by health care professionals is not allowed according to Article 21 of the Medizinprodukteverordnung (MepV, Medical Devices Ordinance). This means that promotion to patients of medical devices is allowed that do not have to be prescribed by health care professionals and that do not have to be applied by professionals.

### 19.3.7. Belgium

In Belgium, promotion of medical devices to patients usually is allowed with the exception of implants! Article 9 (4) of the Belgium law dated March 25, 1964 on medical devices (Loi sur les medicaments de 25 mars 1964) defines that implantable medical devices are not to be promoted to layman. This law defines "implants" as a medical device that is partially or completely implanted in the human body or a device that replaces an epithelial surface or an eye surface via a surgical intervention and that remains there after the intervention.

### 19.3.8. Other EU Countries

In other EU countries the promotion of medical devices is not strictly regulated. It is allowed to address the consumer/patient via promotion. It goes without saying that a medical device manufacturer should only use promotional messages that can be proven with available documents and clinical studies in order to avoid legal problems. This conduct is for example explicitly required by the Irish Medicines Board. When developing the promotional material, medical device manufacturers should comply with the following EU directives:

- Directive 2005/29/EC from May 11, 2005 concerning unfair business-to-consumer commercial practices in the internal market *and*
- Directive 2006/114/EC from December 12, 2006 concerning misleading and comparative advertising

### 19.4. Promotion for Medical Devices without a CE-mark

The promotion for medical devices without a CE-mark is possible at conventions and symposia in Austria, Belgium, Ireland and Germany. The prerequisite is that the promotional material clearly states that the CE-mark has not been granted yet and that no comparison to competitive products is made.

### 19.5. Summary

The different legal regulation status in EU countries makes it difficult for medical device manufacturers to develop their promotional activities to health care professionals, consumers and patients in compliance with the legal regulations. Moreover, they cannot develop one promotion for all target groups but have to adapt it the requirements of the respective EU country.

The prohibition of the promotion of professional medical devices to patients in countries like Italy, Spain, France, Austria and Switzerland is meant to limit the influence of the pharmaceutical industry to patients who are not well informed and not up-to-date.

It would be appreciated if there would be a EU-wide binding legislation regarding the promotion of medical devices and – in countries that limit or prohibit promotion to consumers/patients – a loosening of this legislation.

### 19.6. Test Your Knowledge

| | |
|---|---|
| **Q1:** | What is meant by "professional medical devices"? |
| **A1:** | Professional medical devices are products that can only be prescribed by a physician. Non-professional medical devices are for example contact lens care or a fever thermometer. All kinds of implants are professional medical devices. |

| | |
|---|---|
| **Q2:** | In which countries is the promotion of professional medical devices allowed: In 1. Germany, 2. Switzerland or 3. Belgium? |
| **A2:** | This question can only be answered for 1. Germany. Promotion for professional medical devices to patients is allowed. In Belgium, promotion to patients is generally allowed with the exception of certain implants. In Switzerland only medical devices – that do not have to be prescribed by a physician do not have to be applied by health care professionals – can be promoted to patients. |

### 19.7. References

- Heilmittelwerbegesetz (HWG, German Law on Advertising of Medicinal Products), www.gesetze-im-internet.de/bundesrecht/heilmwerbg/gesamt.pdf
- www.bvmed.de/themen/kommunikation/Werbung_fuer_Medizinprodukte/articl e/Praxisleitfaden_HWG.htm
- www.bvmed.de/themen/kommunikation/Werbung_fuer_Medizinprodukte/articl e/2011-06-kommunikation-und-werbung-fuer-medizinprodukte-was-ist-erlaubt-was-ist-verboten-ein-praxisleitfaden-zum-umgang-mit-dem-heilmittelwerbegesetz-hwg.html
- Directive 2006/114/EC, http://eur-lex.europa.eu/LexUriServ/LexUriServ.do? uri=OJ:L:2006:376:0021:0027:DE:PDF

- www.legifrance.gouv.fr/affichTexte.do?cidTexte=JORFTEXT000026835146&fastPos=47&fastReqId=1710464764&categorieLien=id&oldAction=rechTexte)
- www.legifrance.gouv.fr/affichTexte.do?cidTexte=JORFTEXT000026451423&fastPos=16&fastReqId=1473494667&categorieLien=id&oldAction=rechTexte)

**Chapter 20:** **Market Surveillance and Reporting Obligations** - Market
Surveillance and Vigilance, Post-market Surveillance (PMS) and PMS
Plan, Post- market Clinical Follow-up and Corrective Measures
*Myriam Becker, Dr. Sibylle Scholtz*

20.1. Learning Objective

The safety of medical devices is an essential objective for manufacturers, patients, users and third parties. The European directive as well as national regulations define mandatory requirements. In this chapter you will learn about the requirements derived from MDD 93/42/EEC for manufacturers as well as users of medical devices in order to avoid hazards to patients, users and third parties by the medical device. The national regulations implementing this directive, like the Medizinproduktegesetz (MPG, Medical Devices Act) and the Medizinproduktesicherheits-Planverordnung (MPSV, Medical Device Safety Plan Regulation) in Germany will be discussed in this chapter.

20.2. Basics and Definitions in the European Environment Surveillance

After market launch of a product, the manufacturer has to establish and maintain a system to trace the suitability of the products. This system has to actively monitor, list and assess whether all product properties – defined by the requirements and the technical documentation – are met over the whole product life cycle (Post Market Surveillance System, PMS system). Ongoing monitoring and assessment of the risk-benefit-ratio is key. In case that in the course of time there are more incidents than foreseen, corrective measures must be taken by the manufacturer (www.basg.gv.at/medizinprodukte/vigilanz-und-marktueberwachung).

Worldwide, legislators require that manufacturer of medical devices intensively monitor the market in order to avoid hazards and risks to patients, users and third parties that could be provoked by their medical devices. The basis for market surveillance in Europe is directive 93/42/EEC for medical devices as well as the requirements from 2007/47/EC (http://ec.europa.eu/health/medical-devices/files/ revision_docs/2007-47-en_en.pdf) (English version) and the directive 90/385/EEC on active implantable medical devices (AIMD) (http://eur-lex.europa.eu/LexUriServ/ LexUriServ.do?uri=CONSLEG:1990L0385:20071011:en:PDF) as well as the

directive 98/79/EC on IVDs (http://eur-lex.europa.eu/LexUriServ/LexUriServ.do?uri=
CELEX:31998L0079:en:NOT).

These three directives were summarized in the Medizinproduktegesetz (MPG,
Medical Devices Act) www.gesetze-im-internet.de/bundesrecht/mpg/gesamt.pdf).
This act is complemented by a number of implementing regulations, in this case by
the Medizinprodukte-Sicherheitsplanverordnung (MPSPV, Medical Device Safety
Plan Regulation), on the directive on the collection, assessment and prevention of
risks caused by medical devices (www.gesetze-im-internet.de/bundesrecht/mpsv/
gesamt.pdf). All legal regulations are published on the homepage of the BfArM
(Federal Institute for Medicines and Medical Devices) (http://www.bfarm.de/DE/
Medizinprodukte/rechtlicherRahmen/gesetze/_node.html).


Two aspects have to be taken into consideration regarding the obligatory market
surveillance: on one hand the market surveillance of the own products and on the
other hand the market surveillance of comparable medical devices of competitors. In
the Arzneimittelgesetz (AMG, Medicines Act), the identification of adverse events can
be ensured with a regular literature search. As far as medical devices are concerned
a comparable definition does not exist (not yet). The lack of a central database also
makes it difficult for the manufacturer to comply with the legal requirements. Only in
MEDDEV 2.12/2 literature search is mentioned as a possible way for post-market
clinical follow-up (http://ec.europa.eu/health/medical-devices/files/meddev/2_12_2_ol
_en.pdf).

§ 37 MPG (Medical Devices Act) is the basis for the Medizinprodukte-Sicherheitsplanverordnung (MPSV, Medical Device Safety Plan Regulation). MPSV is
mandatory for all users and operators of medical devices. The MPSV is not applied to
medicinal products that are used during clinical studies or to IVDs for performance
assessments.

---

**Exercise:**

Look for further information on the surveillance of medical devices in the following documents:

MEDDEV 2.12/2: Guidelines on post-market clinical follow-up studies, http://ec.europa.eu/health/medical-devices/files/meddev/2_12_2_ol_en.pdf

NBMed 2.12 rec 1, Post-Marketing Surveillance (PMS) post market / production, www.team-nb.org

---

<u>Vigilance</u>

The term "vigilance" derives from the Latin word "vigilantia" and can be translated with "truth" or "ingenuity". Vigilance together with medical device means the obligation of notification of "incidences" (adverse events) that arose when using the medical device. This "incidence" could be a malfunction, defaults or modification of characteristics or of performance or inappropriate labeling or inappropriate instructions for use of a medical device that lead / may lead or may have led – directly or indirectly – to death or a serious deterioration of health of a patient, a user or another person (http://www.bfarm.de/DE/Medizinprodukte/risikoerfassung/RisikenMelden/_node.html).

According to the requirements of the European directive 93/42/EEC – and in Germany according to the requirements of the MPSV – those who place medical devices on the market (manufacturer, authorized representatives or importer) are responsible to notify the BfArM on incidences that arose in Germany with a product or on conducted recalls of a product. In case of IVDs the Paul Ehrlich Institute (PEI) has to be notified.

That means that every member state is required to systematically collect and assess all reported incidences and to notify the European Commission and the other member states on corrective measures. The vigilance system is an additional safety element to ensure the protection of health as well as to enable the comprehensive, fast implementation of corrective measures and to provide an efficient risk management. This should prevent the same incidences from happening again and adds to the scientific level of knowledge.

The obligation to notify exists also for professional operators and user (f. e. physicians and dentists) and third parties that – professionally or commercially – use or provide medical devices to consumers. The notification forms can be found on the homepage of the BfArM (http://www.bfarm.de/DE/Service/Formulare/functions/ Medizinprodukte/ node.html).

When deciding whether or not an incidence has to be reported, the BVMed decision tree can be helpful:

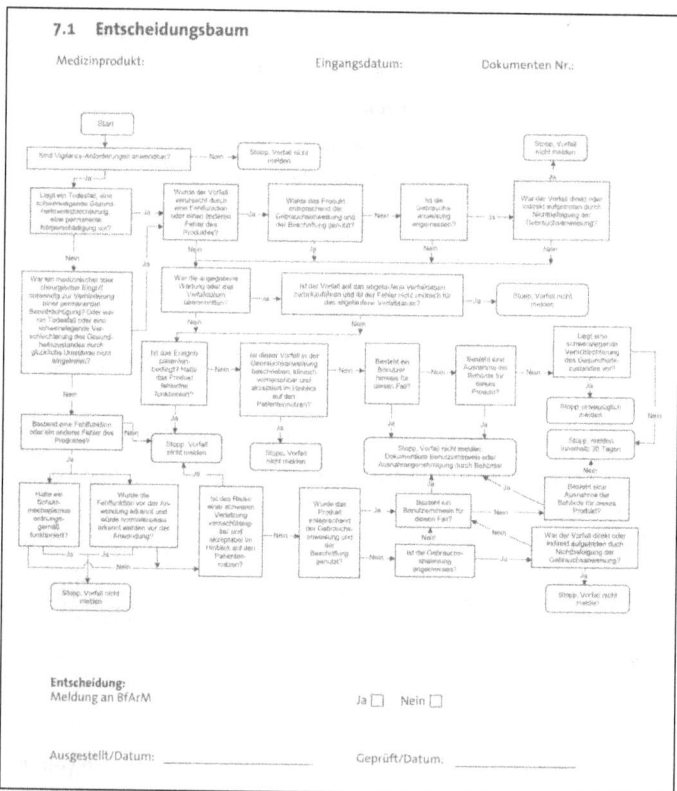

Fig. 20/1: Decision tree for reportable incidents (BVMED, Informationsreihe Medizinprodukterecht, Marktüberwachung von Medizinprodukten, Leitlinie zur Meldung von Vorkommmnissen, BVMED, Berlin, Juni 2012).

If the manufacturer has decided that an incidence has to be reported, the following procedure starts:

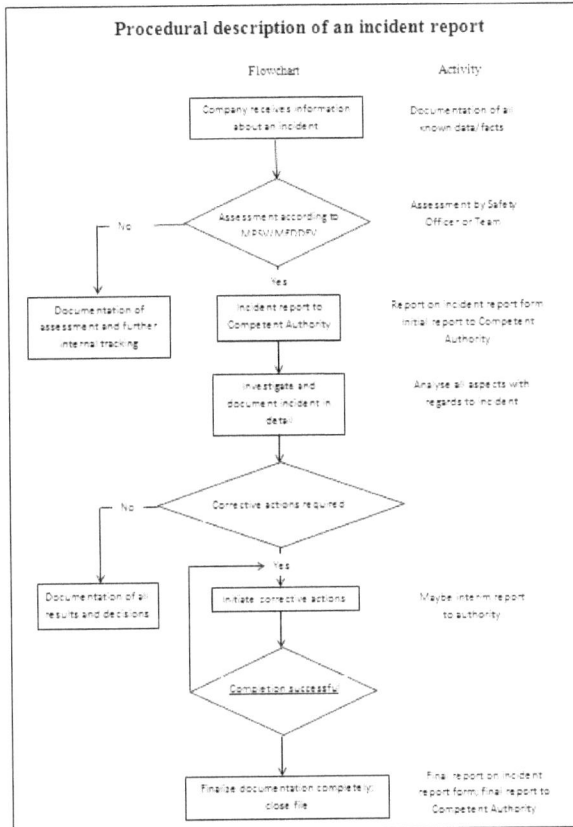

Fig. 20/2: Procedural description of an incident report (created by the authors, based on BVMED, Informationsreihe Medizinprodukterecht, Marktüberwachung von Medizinprodukten, Leitlinie zur Meldung von Vorkommnissen, BVMED, Berlin, 06/2012).

The documentation and administration of the reports in Germany takes place at the Deutsche Institut für Medizinische Dokumentation und Information (DIMDI, German Institute for Medical Documentation and Information) (www.dimdi.de/static/de/ index.html).

Recommendation concerning a vigilance system is provided by the Medicines and Healthcare Products Regulatory Agency (MHRA) on a European level (MEDDEV 2.12/1, GUIDELINES ON A MEDICAL DEVICES VIGILANCE SYSTEM) (www.mhra.gov.uk/index.htm#page=DynamicListMedicines and http://ec.europa.eu/ health/medical-devices/files/meddev/2_12_1_ol_en.pdf).

---

**Exercise:**

Look for further information on the internet, for example MEDDEV 2.12-1, GUIDELINES ON A MEDICAL DEVICES VIGILANCE SYSTEM, http://ec.europa.eu/health/medical-devices/files/meddev/2_12_1_ol_en.pdf. Carefully read the notification forms for incidences with and recalls of medical devices (http://www.bfarm.de/DE/Service/Formulare/functions/Medizinprodukte/_node.html). As examples of national implementation of the European directive 93/42/ECC two countries will be focused on: Germany and Austria.

---

## 20.3. Example for the Implementation into National Law: The German Medical Device Vigilance System

### 20.3.1. Tasks of the Authority

In order to avoid hazard to the health of patients, users and third parties, the higher federal competent authority has to centrally collect and assess all risks that may arise from the use of the medical device. These are risks like side effects, interactions with other substances or products, contraindications, tampering, functional errors, malfunction and technical defects. When serious incidents and adverse events arise, the higher federal authority (BfArM or PEI) has to coordinate specific measures. Special focus is put on serious adverse events during clinical trials or trials evaluating performance on one hand and incidents with medical devices already available on the market on the other hand.

The following events are defined as "serious adverse events" when they arise during clinical trials (subject to approval) or trials evaluating performance: events that lead, have led or might lead - directly or indirectly - to the death or serious deterioration of the health of a test subject without considering whether this incident was caused by the medical device. Every functional impairment, change in characteristics or performance, inadequacy of the labelling or the instructions for use of a distributed medical device that leads, has led or might have led to the death or a serious deterioration of health of patients, is called an "incident". A serious deterioration of the health is a life-threatening disease or harm, a permanent impairment or injury of a body function or in case that medical or surgical interventions were necessary to avoid this from happening. In this context, a product recall can be ordered by the authorities. This kind of recall is a corrective measure that requires the return, the exchange, the refitting or retrofitting or even the disposal of a medical device. Moreover, users, operators or patients get further instructions for a safe use of the medical device. A corrective measure leads to the reduction, elimination and/or prevention of recurrence of the defined incident/risk.

With corrective measures, the following documentation needs to be submitted:

- Relevant parts of the risk analysis
- Background and justification for corrective measures incl. the description of the problem
- Risks that are linked to the ongoing use of the product
- Risks for patients that already used or are currently using the product
- Solution of the problem affected products (serial number, LOT number)
- Clear identification of the manufacturer or his European representative
- Basic guideline for the distributor and/or user, e.g. do not use product anymore, dispose of product
- Instructions regarding follow-up of affected patients
- Instructions to distribute the information within the organisation to everybody who is dealing with the affected product
- Keeping the information sufficiently long
- In case that potentially affected products were distributed to third parties: to inform the third parties or the manufacturer
- Content of the customer information with corrective measures

- Urgent recommendation
- Title (field safety corrective action)
- Clear identification of the product
- "Reason why" for this customer information incl. description of the problem and the risk arising from the product
- Clear recommendation of measures to the user; no advertising for products or service
- Contact person
- Instruction to keep the information long enough - until the measure is concluded
- Instruction that the Federal Institute for Medicines and Medical Devices received a copy of this "urgent safety information".

One has to make sure that the content is approved by the competent authority of the manufacturer/representative (deadline for comments: 48 h). A copy of the customer information should be submitted to the Notified Body (BVmed, Informationsreihe Medizinprodukterecht, Marktüberwachung von Medizinprodukten, Leitlinie zur Meldung von Vorkommnissen, BVMED, Berlin, June 2012 und BfArM, http://www.bfarm.de/SharedDocs/Formulare/DE/Medizinprodukte/VorlageHersteller.p df?_blob=publicationFile&v=3

The basis for this is that the responsible person according to § 5 MPG (Medical Devices Act), e. g. the manufacturer) assumes the responsibility for the medical device and focuses in his or her own interest on the clarification of the incidents and on the implementation of corrective measures on their own responsibility.

Operators as well as users have an essential role in the communication and cooperation in investigating and clarifying incidents. The competent higher federal authority (BfArM, PEI) has the role of assessing and scientifically evaluating the incidents and of coordinating the measures. The competent state authorities have a controlling and steering function. The state authority also may order measures to be implemented or conduct own investigations.

20.3.2. Reporting Obligation

With reference to §3 MPSV, manufacturers, operators, users or distributors of medical devices within the European Economic Area and in Switzerland are obliged to report incidents with CE-marked products as well as Products without a CE-mark that are within the scope of the directive (e. g. custom-made devices). The notification criterion is that the incident has already occurred and that the medical device of the manufacturer has contributed or might have contributed to it (contrary to incidents within clinical trials) and that the incident has led or might lead to death or a serious deterioration of the health of a patient, user or third party.

Life-threatening disease or harm is defined: as a permanent impairment of a body function or a permanent damage to the body structure or a condition that leads to a medical or surgical intervention in order to avoid a permanent impairment of a body function or a permanent damage to the body structure, a significant increase of the operation time, necessary treatment in a hospital or an extension of an ongoing hospital treatment, unnecessary or false treatment due to incorrect diagnostics, risk exposure of an unborn child, its death or any anomaly or birth defects.

Manufacturers have to report incidents in Germany, systematically conducted recalls in Germany as well as incidents and recalls outside the European Economic Area to BfArM or PEI and also in case the recall takes place in the European Economic Area (§ 32 MPG, tasks and responsibilities of the higher federal authorities regarding medical devices). Generally speaking, the manufacturer of a medical device is obliged to inform the competent authority of a European country on the incidents and product recalls that were reported/conducted in this country. Moreover, the manufacturer has to inform his Notified Body on problems that might affect the CE certificate. Also, if a manufacturer is exempt from the obligation notification by the authority, the manufacturer still has to investigate and assess the incidents. The competent authority has to inform the manufacturer on reports sent by the authority (BVMED, Informationsreihe Medizinprodukterecht, Marktüberwachung von Medizinprodukten, Leitlinie zur Meldung von Vorkommnissen, BVMED, Berlin, June 2012).

If during a treatment by operators and users, incidents with medical devices – that were applied to patients – become known, these incidents also have to be reported to the BfArM and PEI. Distributors (professional, commercial or who are fulfilling legal objectives or responsibilities) of medical devices for self-application by patients and other laymen, are also obliged to report such incidents.

According to § 4 MPSV there is an exemption of the notification obligation in case a) these incidents are already adequately investigated and b) whether corrective measures have already been conducted:

- Device errors that were identified before the use
- Problems that are attributable to the patient or his disease
- Transgression of the defined life cycle of a medical device
- First-fault safety did work (alerts ...)
- Very low occurrence probability of death or a serious deterioration of health
- Expected and foreseeable side effects (instructions for use!)

Also, use errors have to be reported, in case

- They have led to death or a serious deterioration of health
- They have led to corrective measures in order to avoid death or a serious deterioration of health
- They prevent a significant increase of the frequency of use errors that are potentially fatal or very serious.

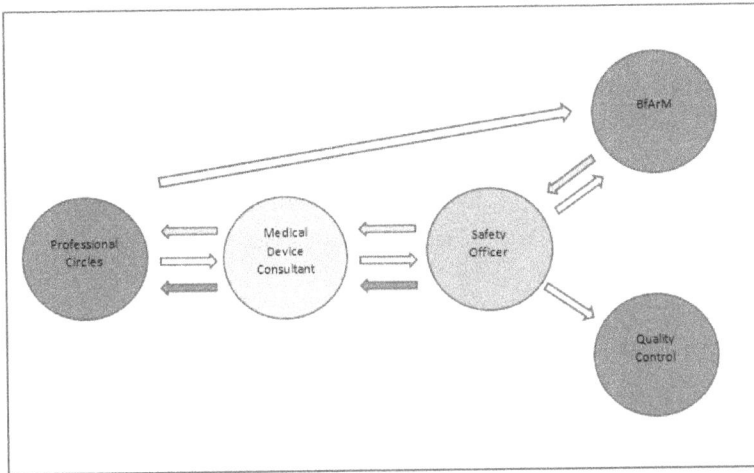

Fig. 20/3: The reporting system of professional circles, medical device consultants, safety officers and the BfArM at a glance (created by the authors, based on BVMED, Informationsreihe Medizinprodukterecht, § 31 Medizinprodukteberater, Teil 1, Basismodul, BVMED, Berlin, Januar 2008).

## 20.3.3. Notification Deadline

In § 5 MPSV the notification deadlines are defined: in general, the notification takes place regarding its urgency. In case of a serious public health threat, the notification must take place immediately, at the latest within 2 calendar days after becoming aware of the hazard. "Immediately" is defined as "without delay that cannot be justified". In case of imminent danger, the notification has to take place immediately. In case of death or unforeseeable serious deterioration of health, the manufacturer has to notify immediately after becoming aware of the incident and ensuring the participation of his product, at the latest after 10 calendar days.

"Unforeseeable" means that the manufacturer has not considered the problem in his risk analysis. In all other cases, the notification has to take place immediately after becoming aware of the incident and ensuring the participation of the product, at the latest after 30 calendar days. Reports by operators, users and traders have to be reported immediately. Recalls on the basis of "mandatory reportable incidents" outside the EU have to be reported immediately by the manufacturer, at the latest when corrective measures are being implemented.

231

§3 MPSV also defines notification deadlines that have to be considered when serious events arise during a clinical trial. If the adverse event arises in Germany, the sponsor of the clinical trial has to notify the BfArM or PEI who are the principal investigators. In addition to this, the competent authority has to be notified in case the clinical trial is conducted in this country. If the serious event arises outside of Germany, the sponsor also has to notify the competent authority of the respective EU country, and only the BfArM or PEI if the trial is also conducted in Germany.

### 20.3.4. Notification Procedure

According to § 7 MPSV and the announcement of the German Ministry of Health of June 6, 2002 notifications are transferred electronically. The applicable forms are published on the website of the BfArM (http://www.bfarm.de/DE/Service/Formulare/ functions/Medizinprodukte/_node.html) and DIMDI (www.dimdi.de).

Special attention is paid to implants. Operators have to establish and keep records according to §16 MPSV on the name, date of birth and address of a patient, date of the implantation type and serial number as well as LOT number of the implant as well as the name of the responsible person according to § 5 MPG (Medical Devices Act) up to 20 years after the implantation. This includes active implantable devices such as cardiac pacemakers, defibrillators or infusion systems and also for other implantable medical devices such as heart valves, vascular prosthesis and stents, breast implants, hip prosthesis and endoprosthesis.

### 20.3.5. National Specificities in Germany and Austria: Safety Officer and MD Consultants

When drafting the national MPG (Medical Devices Act), Germany and Austria implemented two additional functions: the function of a "medical device safety officer" and the function of a "MD consultant".

The role of a safety officer is described in § 30 MPG (Germany) and in § 78 MPG (Austria): Everyone who is responsible for placing medical devices on the market for the first time has to appoint a safety officer according to the MPG (Medical Devices Act). The safety officer has to have the required expertise and reliability to confidently handle legal requirements and should be adequately integrated inside the company in order to avoid conflicts of interests (e. g. no Sales or Marketing representatives). Upon assuming his or her role, the competent authority must be informed. The required expertise is defined by being a holder of a university degree (scientific, medical or technical) or a comparable education that enables the safety officer to exercise his tasks and at least a 2-year working experience.

The safety officer has to collect reports on risks linked to the medical device, to assess these reports and to coordinate necessary measures as well as to notify the competent authority. The fulfilment of his or her assigned tasks should not be of disadvantage to the safety officer and he or she must be able to perform the assigned tasks on his or her own responsibility.

The tasks of a MD consultant consist of professional informing and consulting the relevant professional circles and to introduce them to the proper handling of the medical devices. Relevant professional circles are health care professionals, health care facilities staff and persons who manufacture, assess, place products on the market, implant devices, put devices into operation, operate and apply devices (§ 31 MPG Germany, § 79 MPG Austria). The law does not provide for the information of patients and consumers. In general, the MD consultant is a self-employed commercial agent or an employee of a MD manufacturer for which he places the products on the market.

20.3.6. Obligation of the Manufacturer: Product Surveillance

The surveillance of the product is also partly an obligation of the medical device consultant. New information resulting from MD consultant feedback and from scientific standard literature and the media are provided to the competent authority. Depending on the risk potential of the medical device, the following may be necessary: extensive literature searches, additional tests of the MD and additional technical studies, as well as clinical follow-up trials after market entry, the systematic compilation, assessment and evaluation of user experience and information from implants registers in the context of market surveillance.

### 20.3.7. Tasks of the Higher Federal Authority

Then higher federal authorities BfArM and PEI have to collect, assess and evaluate incidents and coordinate measures in Germany. BfArM and PEI are getting into action if the incidents take place in Germany and if the manufacturer resides in Germany. The information is addressed to the competent authority of the manufacturer and to the state/county authority that is responsible for the place where the incident happened.

The risk assessment of incidents is carried out by the BfArM or PEI in cooperation with the manufacturer and possibly with the competent authority of the manufacturer. Also, operators and users may be called on. The one who is responsible for placing the product on the market has a legally defined obligation to actively participate in the risk assessment – as well as the operator.

If the BfArM or PEI come to the conclusion that it was an application error or an isolated case, no corrective measures are necessary. After informing the responsible person, the reporting person and the competent authority, the processing of the matter will be completed by the BfArM or PEI. In all other cases, corrective measures are necessary. If there is mutual agreement with the responsible person, the responsible person is informed, as well as the reporting person and the local competent authority (that might impose measures!). After that, the processing of the incident is finalised by the BfArM or PEI. If a corrective measure cannot be agreed upon with the responsible person, the procedure is the same: after informing the responsible person, the reporting person and the competent (regional) authority that is responsible for the measures (authority of the manufacturer: measures!), the processing of the incident is completed by the BfArM or PEI.

The risk assessment of the competent local authority usually follows the assessment of the BfArM or PEI. Ideally, the manufacturer comes to a decision on his own responsibility rather than waiting for state actions.

### 20.3.8. Official Measures

In specific situations the local authority will determine necessary measures in ascending escalation order. First of all, the manufacturer starts monitoring his measures. Official measures of level I are – where appropriate – the active monitoring of the implementation of corrective measures of the manufacturer. At the next escalation level (level II) the authorities responsible for the manufacturer impose measures that the manufacturer has to comply with in case the manufacturer has not implemented mutually agreed corrective measures or did not see the need for them.

At the last escalation level (level III) the local competent authority imposes measures that are mandatory to the operator or user in case the manufacturer did not implement the mutually agreed corrective measures or did not see the need for them and in case the operation of the medical device can create a hazard.

### 20.3.9. Possible Consequences of Incidents

In order to be able to use new information gained from incidents in an appropriate way to improve a medical device and to avoid recurrence of these incidents (that means in order to protect patients, users and third parties), the legislator has defined a number of possible consequences:

The first priority is the evaluation of the incident and the decision regarding the obligation of notification. The direct and periodic assessment of the risk profile and if appropriate its actualization may also be a possible consequence as well as correction of the product labeling (e. g. addition of warnings) and necessary trainings of the staff of the MD manufacturer as well as of the users. Moreover, a re-design of the affected product might be necessary in order to be able to further distribute the product.

If a manufacturer refers to the MDD 93/42/EEC, post-market surveillance and post-market clinical follow-up becomes mandatory. Moreover, it is an obligation to keep the product "state of the art". A screening of the literature is part of the post-market surveillance, of the clinical assessment as well as of the post-market clinical follow-up. For the clinical assessment of medical devices of all classes (according to 2007/47/EC) the essential requirements have to be met (safety & effectiveness, beneficial risk/benefit ratio).

There are also possible consequences in reference to the clinical assessment and risk assessment according to MDD, Annex X, 1.1.c: In this section, the necessity of an active actualization of the clinical assessment based on post-market surveillance data is described. The MDD considers the actualization of the risk assessment as necessary as the clinical assessment at defined intervals, the taking into account of PMS data (incl. incidents and literature) and the necessity of a PMS plan for every product family.

### 20.4. Requirements of the Quality System (MDD, Annex II, 3):

Known risks (e. g. from literature or from incidents) are included in the risk management process and have to be addressed in the clinical assessment. Residual risks have to be shown in the labeling. The medical benefit has to be bigger than the residual risk (risk-benefit-ratio).

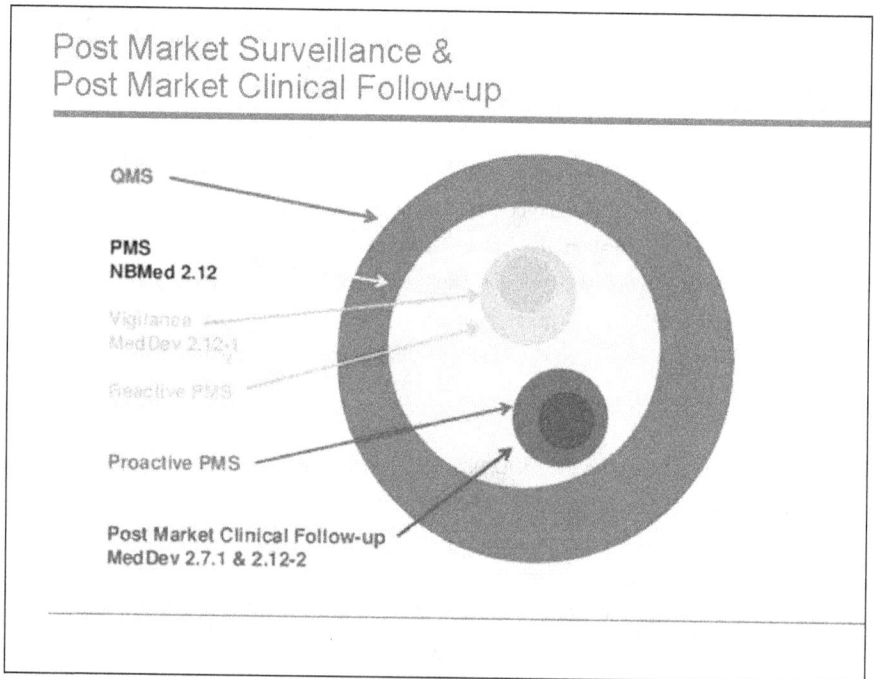

## Post Market Surveillance & Post Market Clinical Follow-up

QMS

PMS
NBMed 2.12

Vigilance
MedDev 2.12-1

Reactive PMS

Proactive PMS

Post Market Clinical Follow-up
MedDev 2.7.1 & 2.12-2

Fig. 20/4: Mapping of specific processes in Post Market Surveillance and Post Market Clinical Follow-Up (own illustration)

## 20.5. Statistics

Updated numbers of incidents of medical devices in Germany are published on the website of the BfArM (http://www.bfarm.de/DE/Medizinprodukte/risikoerfassung/wissauf/ _node.html).

---

**Exercise:**

Go to the BfArM website and look for details on causes of error (e. g. design and construction errors, production errors, other product-related causes or non-product related causes).

---

## 20.6. Summary

An efficient market surveillance of medical devices is in the interest of patients, users and third parties who want to or have to trust in the safety of the applied products. It is also of high importance for the competent authorities and especially for the manufacturer. Market surveillance is the basis for reliably reporting incidents and for processing of these incidents by all responsible entities.

Medical devices that have undergone the conformity assessment procedure are considered as safe, efficient and therapeutically useful. An adequate and economic useful market surveillance is necessary to improve patient safety and optimized medical care of the patients with medical devices.

The objective of the planned evolution of market surveillance is establishing unique, quality-assured market surveillance by the Bundesländer (federal states). The draft defines for example the taking of samples, the procedures following detected deficiencies or quality assurance measures of the competent authority and a better exchange of information between the authorities. Current European activities are for example a regulation on electronic instructions for use of medical devices, the modification of the EU directive on IVDs as well as a regulation on special requirements of medical devices incorporating material of animal origin (TSE/BSE).

The discussion focuses on the review of the European legal framework for medical devices. The existing directives shall be transferred into regulations and thus become directly applicable law. The objective is to guarantee a high level of safety and health protection – especially in the light of new technological developments as well as to guarantee an innovation-friendly legal framework that enables a quick market entry. There are a lot of open questions and various alternative options, for example as far as the monitoring of the work of the Notified Bodies is concerned. The EU commission also discusses an option to introduce an official control of the conformity assessment by Notified Bodies for special high-risks products or technical innovations. As far as clinical trials are concerned, an obligation of the manufacturer to prove the "medical benefit" of the device is being considered. Regarding the reprocessing of single-use products two options are currently discussed by the commission: a ban or the definition of requirements that impose the obligations of a manufacturer on the companies reprocessing these devices.

The name Eudamed – the European database project – stands for **European Data**bank on **M**edical **D**evices. The development of Eudamed started with the first pilot project of the Deutsche Institut für Medizinische Information und Dokumentation (DIMDI, German Institute for Medical Information and Documentation) in 1999. Since 2000 the Commission is in charge of Eudamed.

The database shall enable the exchange of information on the following topics:

- Notification by manufacturers, representatives and medical devices of lower classes
- Certificates of Notified Bodies
- Surveillance and notification procedure

The usage of Eudamed is mandatory since spring 2011. This system offers a central data exchange on medical devices between the competent authorities of the member states. The database is not publicly accessable. Eudamed consists of 5 modules: acting individual (for example the manufacturer), products, certificates, surveillance and notification procedures as well as clinical trials. At the moment the project does not directly influence manufacturers. Indirect influence is possible via adjustments of the national notification systems. Existing national notification obligations are not affected by Eudamed. In the future, Eudamed could become a central, publicly accessible portal for reporting by manufacturers, representatives and medical devices (also see BVMed-Konferenz zum Medizinprodukterecht: "Regulatorisches

Umfeld der MedTech-Branche ändert sich in den nächsten Jahren, 10. November 2011, Bonn (http://www.bvmed.de/themen/regulatory/Medizinprodukterecht MPG/pressemitteilu ng/bvmed-konferenz-zum-medizinprodukterecht-regulatorisches-umfeld-der-medtech-branche-aendert-sich-in-den-naechsten-jahren.html).

## 20.7. Test Your Knowledge

**Question:**

Market surveillance is mandatory for...?

**Answer:**

... the manufacturer of a medical device.

---

**Question:**

What notification deadline is defined by § 5 MPSV?

**Answer:**

Immediately (imminent danger), at the latest within 30 calendar days.

---

**Question:**

What fine (max) can be imposed on a MD consultant if he does not report an incident in the defined form?

**Answer:**

25.000 Euro

## 20.8. References

- MDD 93/42/EEC (http://ec.europa.eu/health/medical-devices/files/revision docs/2007-47-en en.pdf (English) and http://eur-lex.europa.eu/LexUriServ/ LexUriServ.do?uri=OJ:L:2007:247:0021:0055:de:PDF (German))

- MPG (Medical Devices Act) www.gesetze-im-internet.de/bundesrecht/mpg/ gesamt.pdf

- MPSV (Medical Device Safety Plan Regulation): Regulation on the collection, assessment and prevention of risks of medical devices (www.gesetze-im-internet.de/bundesrecht/mpsv/gesamt.pdf)

- www.basg.gv.at/medizinprodukte/vigilanz-und-marktueberwachung)

- Directive 2007/47/EC amending Council Directive 90/385/EEC on the approximation of the laws of the Member States relating to the active implantable medical devices, Council Directive 93/42/EEC concerning medical devices and Directive 98/8/EC concerning the placing of biocidal products on the market

- Directive 90/385/EEC on the approximation of the laws of the Member States relating to the active implantable medical devices

- (http://eur-lex.europa.eu/LexUriServ/LexUriServ.do?uri=CONSLEG: 1990L0385:20071011:en:PDF)

- Directive 98/79/EC on IVDs (http://eur-lex.europa.eu/LexUriServ/LexUriServ. do?uri=CELEX:31998L0079:en:NOT)

- www.bfarm.de/DE/Medizinprodukte/mpRecht/mprecht-node.html

- MEDDEV 2.12/2 POST MARKET CLINICAL FOLLOW-UP STUDIES: A GUIDE FOR MANUFACTURERS AND NOTIFIED BODIES (http://ec.europa.eu/health/medical-devices/files/meddev/2_12_2_ol_en.pdf)

- www.team-nb.org

- http://www.bfarm.de/DE/Medizinprodukte/risikoerfassung/RisikenMelden/_nod e.html

- http://www.bfarm.de/DE/Service/Formulare/functions/Medizinprodukte/_node.h tml

- BVMED information series on Medical Devices Act, market surveillance of MDs, Directive on the notification of incidents, BVMED, Berlin, June 2012

- www.dimdi.de/static/de/index.html

- MEDDEV 2.12/1, GUIDELINES ON A MEDICAL DEVICES VIGILANCE SYSTEM (http://ec.europa.eu/health/medical-devices/files/meddev/2_12_1_ol _en.pdf)

- www.mhra.gov.uk/index.htm#page=DynamicListMedicines

- www.bfarm.de/DE/Medizinprodukte/vigilanz/vigilanz- node.html#doc1012384bodyText1

- BVMED information series on Medical Devices Act, § 31 MD consultant, part 1, BVMED, Berlin, January 2008

- MPG (Medical Devices Act), Austria www.bmgfj.gv.at/cms/home/attachments/7/9/0/CH1095/CMS1228212263945/ medizinproduktegesetz_kompiliert.pdf

- http://www.bfarm.de/DE/Medizinprodukte/risikoerfassung/wissauf/_node.html
- BVMed conference on the Medical Devices Act „Regulatory environment of the MedTech industry will change within the next years", November 10, 2011. http://www.bvmed.de/themen/regulatory/Medizinprodukterecht_MPG/presse mitteilung/bvmed-konferenz-zum-medizinprodukterecht-regulatorisches-umfeld-der-medtech-branche-aendert-sich-in-den-naechsten-jahren.html

Market Surveillance and Reporting Obligations

**Chapter 21: Change Management / Lifecycle Management / Product Modification**

*Dr. Stefan Menzl*

## 21.1. Learning Objective

„Nothing is as constant as change" already stated Heraclitus of Ephesus (*520 BC; †
460 BC).
The same is true for medical devices. In this chapter you will learn about the
regulatory activities or consequences that are relevant in each phase of the lifecycle
of a medical device.

## 21.2. Introduction

Compared to medicines, the lifecycle of medical devices is relatively short. On the
day a medical device is placed on the market it is already likely that it will soon be
replaced by a new version of it.
MD manufacturers navigate in a highly regulated environment. This chapter focuses
on the unavoidable modifications of medical devices and explains which
modifications have to be reported and what other consequences may arise from
modifications. The author uses the example of intraocular lenses to illustrate these
changes.

## 21.3. Lifecycle of a Medical Device

At the beginning of a product lifecycle, there is the idea and the design of a medical
device. A prototype will be developed that will undergo a series of tests in the
preclinical phase. Later, in the clinical phase the device has to poof that it really
meets the specified clinical performance criteria.
The product is finally manufactured, promoted and placed on the market. Only in this
intensive phase of application does the product ultimately prove its technical and
clinical performance in the everyday life. This also is the phase in which a lot of
additional knowhow and inspiration for further development is gained. Sometimes,
this phase also results in the somewhat painful realization that a product does not
meet all necessary requirements and that it has to be modified and shortcomings
addressed, or worst case that the product must even be taken off the market again.
Eventually, a product reaches the end of its lifecycle and is replaced by the next
product generation.

21.4. Reasons for Modification

What triggers modification of an approved product or of its production, labelling or indication? What promotes further development of products?

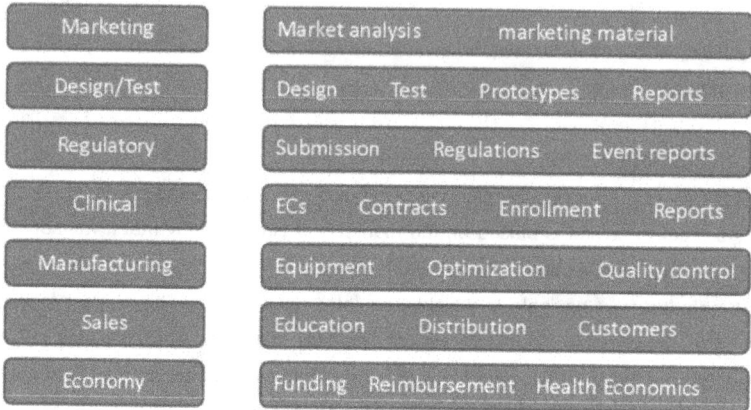

| Marketing | Market analysis | marketing material | |
|-----------|-----------------|--------------------|---|
| Design/Test | Design | Test | Prototypes | Reports |
| Regulatory | Submission | Regulations | Event reports |
| Clinical | ECs | Contracts | Enrollment | Reports |
| Manufacturing | Equipment | Optimization | Quality control |
| Sales | Education | Distribution | Customers |
| Economy | Funding | Reimbursement | Health Economics |

Fig. 21/1: "Regularory Compliance" (Menzl, Stefan: Regulatory Compliance bei Medizinprodukten' FORUM Seminar 2011)

Triggers are often identified in market analysis or when assessing the promotional materials of competitors. Target-users might have other needs than originally expected or competitors claim product properties that are highly appreciated by the users.

Also, approval prerequisites concerning design and safety as well as constantly evolving standards are frequent causes for modification of a product or its production. If, for example, safety standards change in order to take account of the current state of the art, the manufacturer has no choice but to adapt his product to the new requirements.

Moreover, reported experience by using a product can result in the necessity to modify the product. Also, new requirements in the regulatory environment, for example trend reporting of adverse events can trigger product modifications.

Another important source of adaptions is the optimization of manufacturing processes. Cost reduction or the increase in product quality (for example by automation or semi-automation) can be key factors.

Aspects of health economics may also result in product modifications. A company in general will only keep products on the market that are reimbursed by payers.

Selling expenses that are increased by the necessity of temperature-controlled transportation play a role. If it is possible to modify a product in such a way that temperature-controlled transportation is no longer needed, this can have decisive trade and cost benefits.

In the end it always is the customer or user who will decide on success or failure of a product as well as the modification or displacement from the market.


## 21.5. Product Development / Product Lifecycle – Exemplary Description

By describing the design and development of the intraocular lens (IOL, an artificial lens that is implanted during a cataract surgery, a medical device) the triggers and consequences of this evolution are demonstrated.

Cataract is a disease that results in the opacity of the natural crystal lens. In the course of the disease visual impairment will occur gradually which may result in the complete loss of the eyesight. Cataract is the most frequent cause of blindness, 90% of cataracts occur in old age (Scholtz S, Vom Lesestein zur LASIK, Aachen 2006).

Today, about 600.000 cataract surgeries are performed per year in Germany. The first "cataract surgeries" took place about 500 BC by using a (non-sterile) needle to push the opaque lens into the vitreous body of the posterior segment of the eye. Only much later it was completely removed from the eye.

In 1795, the Italian Casaamata implanted glass lenses for the first time that were meant to take over the function of the natural lens. Because of the high weight of a glass lens, this attempt was not successful. At this time, surgeons kept pushing the opaque lens into the vitreous body of the posterior segment of the eye or removed it from the eye thus enabling the patient to at least make out the outlines of objects.

With time, the surgery techniques and hygiene standards were further developed: the incision became smaller, the so-called phaco techniques (fragmentation of the opaque and hardened lens by ultrasonic energy and removal of these fragments) was developed. By using sterile procedures, post-operative infection rates were reduced and by using local anaesthesia, the surgery became bearable for patients.

A quantum leap was taken by Sir Harald Ridley, an English eye surgeon, who noticed – after seeing Royal Air Force pilots of World War II with pieces of shattered canopies in their eyes – that the acrylic plastic material did not trigger any inflammatory reactions or foreign body reaction. This led to the realization that PMMA was an ideal material for IOLs.

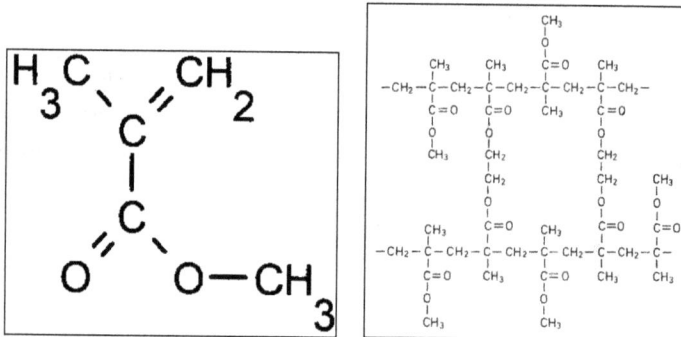

Abb. 21/2, 21/3: left side: MMA (Methylmethacrylat), base for acrylic-IOLs made of PMMA (Polymethylmethacrylat); right side: structure of PMMA

Following years of research by Ridley and the English company Rayner, Ridley implanted the first PMMA IOL on November 29, 1949 into the eye of a 45 year-old cataract patient at the St. Thomas Hospital in London.

It took until the 70s of the last century for the IOL implementation after a cataract surgery to become a standardized treatment.

But complications were inevitable. PMMA is a non-flexible / hard material and therefore cannot be folded / bent. Thus, the incision in the eye had to be quite big (= size of the lens). The consequences of such a big incision were long-lasting wound healing, maybe infections and astigmatism induced by the operation that was perceived as annoying.

Fig. 21/4: Properties of PMMA as IOL material (scientific poster presented at ESCRS congress 2006, Scholtz, Sibylle / Weber, Klaus: From London around the World – Sir Harold Ridley and his idea of Intraocular Lenses conquered the world)

The next milestone for the development of IOLs was triggered by the search for an appropriate material. The objective was to design a more flexible, foldable IOL. About 40 years later, in 1984, the first foldable silicone IOL (suitable for mass production) was placed on the market and successfully implanted.

From then on, cataract surgery followed by IOL implantation became the most frequently conducted surgical intervention worldwide.

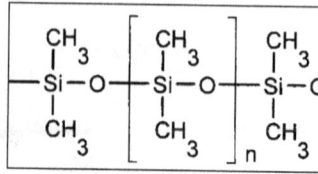

Abb. 21/5: Polydimethylsiloxan

The first IOL materials from silicone were "polydimethylsiloxane". By adding of phenyl groups to the already known molecular structure, a higher refractive index of the material was achieved. This made the production of even thinner IOLs possible. In order to stabilize the material silicates were used.

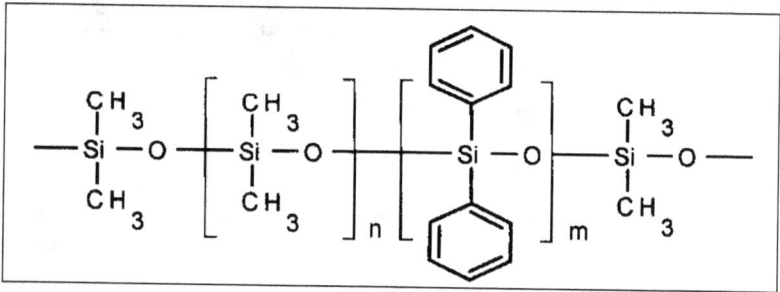

Fig. 21/6: Polydimethyldiphenylsiloxan

The major advantage of silicone lenses is about the flexibility and foldability. This resulted in the implementation via small incisions. The evolution of this material group enabled high-quality products with fewer side effects.

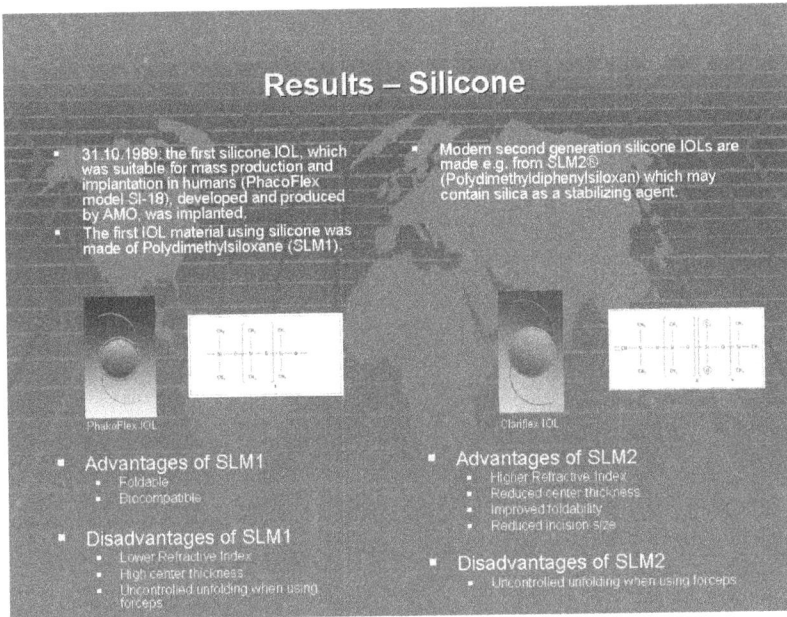

Fig. 21/7: Properties of Silicone as material for IOLs (scientific poster presented at ESCRS congress 2006, Scholtz, Sibylle / Weber, Klaus: From London around the World – Sir Harold Ridley and his idea of Intraocular Lenses conquered the world)

In a next step, the researchers focused on the development of appropriate surgical instruments in order to be able to implant the IOLs in a gentle way with fewer infections. So-called "unfolders" (implementation tools, syringe) with sterile cartridges were used instead of forceps.

As far as the material was concerned, the research went on. Acrylics, the basic component of PMMA were developed further. The substitution of the side chain resulted in a high flexibility of the up to then rigid acrylic material. This resulted in the development of hydrophilic as well as hydrophobic acrylics.

Fig. 21/8: Properties of hydrophobic acrylate as material for IOLs (scientific poster presented at ESCRS congress 2006, Scholtz, Sibylle / Weber, Klaus: From London around the World – Sir Harold Ridley and his idea of Intraocular Lenses conquered the world)

The next development phase focused on the design of the IOL. Starting with the three-piece IOL, consisting of 2 haptics (to stabilize the lens in the capsular bag) and the optic, a one-piece IOL was developed. This resulted in an easy handling of the IOL in the operating theatre and also led to a more cost-effective production.

The next design improvement was the "sharp edge" of the IOL that prevented tissue from migrating behind the IOL after the operation that could lead to lenticular opacity (Posterior Capsular Opacification).

The center-thickness of optical lenses (modern IOLs are much thinner than older IOL types) and their edge design (in order to reduce scatter light effects) were further optimized.

The next milestone of the IOL design optimization was the so-called multifocal lenses. These IOLs enable – like bifocals – a sharp vision in all distances.

The most recent development is the so-called "accommodating" IOL. The ability of the IOL to accommodate is linked to the contraction of the ciliary muscle. This contraction leads to the movement of the IOL or – depending on the IOL model – to the deformation of the IOL. Therefore, these IOLs come very close to the functioning of the natural lens of the eye.

This example shows that there may be developments in the design of medical devices or even completely new MDs might be developed. This may result in significant consequences for the approval of the MD.

Once the decision is taken to modify a product or the manufacturing process, the question then arises whether or not these modifications have to be reported or even authorised.
The NB recommendation NB-MED/2.5.2/Rec2 provides guidance on this topic. This recommendation also contains the requirement that every manufacturer should establish & document a process that defines all changes of the design, product or quality system that require notification AND a process for assessing reporting or authorisation obligations.

---

**Exercise:**
**Go to the MEDDEV website and look for the NB-MED recommendations. Read these carefully!**

---

21.6. Types of Modifications

Generally, modifications can be divided into 2 categories:

1. Modification of the product: these are changes that are linked to the conformity with the essential requirements (MDD) and/or conditions of the intended use.

2. Modification of the quality system which my impact the conformity of the medical device with the essential requirements (MDD).

In addition one differentiates between significant and non-significant modifications, as well as between substantial and non-substantial ones.

Significant changes relate to changes of the product. Substantial changes relate to the quality system or the plans to modify the already approved product line.

The following examples are examples of potentially significant product modifications:
- New risks that have not been considered so far
- Changes that negatively influence the risk assessment
- Changes concerning the intended use
- Changes that are reflected in the list of the essential requirements
- Changes in the user profile
- Change in type of application
- Changes not (yet) covered by the clinical assessment
- Changes of product characteristics or of product performance

The following examples are examples of potentially substantial QS modifications:
- Change in technology
- Change in the product line
- Changes concerning the conformity with the essential requirements

- Changes concerning the conformity with harmonized standards
  - Validation
  - Sterilisation
  - Material
  - Verification
  - Organisational structure
  - Suppliers

## 21.7. Change Notification

As far as significant or substantial changes are concerned, the Notified Body has to be informed. The notification should contain of the following elements:
- Description of the changes as well as a comparison to the CE-marked variant. A tabular listing is recommended.

- Justification of the changes (e.g. changes resulting from a safety-relevant adverse event)
- In case of design/product changes: definition of the relevance relating to the compliance with the essential requirements.

Depending on the Annex of the Directive 93/42/EEC according to which a medical device was granted the CE mark, there are different notification obligations towards the Notified Body.

Examples for significant modifications:

- Changes in material, potential consequences concerning biocompatibility
- Changes in the sterilisation cycle, potential consequences concerning sterilisation validation or sterility
- Modification of the catheter diameter: potential consequences concerning flow and performance
- Changes in packaging configuration: potential consequences concerning the protection of the product during transportation, potential consequences concerning sterility and durability
- Changes in design resulting in modified specifications, potential consequences concerning the essential requirements
- New indication of a product

Examples for substantial (= notifiable) changes of the quality system:

- Changes in the manufacturing process, e.g.application of new technologies
- Exchange of existing design control procedures by new design control proceduresPurchase of a product design that does relate to the approved product line of the manufacturer but which was not developed under his design control procedures
- New clean room, significant extension of an existing clean room or change in clean room class
- Change in the environmental monitoring programme or in the relating control systems
- Change of the sterilising agent or of the sterilisation process (e.g. from EtO sterilization to gamma sterilization)

- Change of the contracting company for sterilization
- Transfer of a product line: from in house to external or vice versa or within the own building

It is often not easy to assess whether modifications are substantial or not. The following questions can be helpful in this assessment:
- Is it a new manufacturing technology?
- Does the modification relate to a product line (extension)?
- Does the change affect the conformity of the product (essential requirements)? Certified design/design type
- Does the modification affect the compliance of the quality system with relevant harmonized standards?
- Does the modification affect major compliance aspects concerning the directive? Verification, validation, organisational structure

If one if the above questions has to be answered with YES, the probability that it is a substantial change is relatively high. An informal discussion with the Notified Body is helpful in any case.

### 21.8. Modifications that are Subject to Approval

Depending on the annex of the Directive 93/42/EEC – according to which a medical device was granted the CE mark – and depending on the risk class of a product, there are different approval requirements. These will be discussed in the following.

Cave: This list is meant to provide orientation and is not intended to be an exhaustive list! In case of doubt, always contact your Notified Body for clarification!

*Medical device, class I, sterile or with measurement function (93/42/EEC)*
*Medical device, class I, sterile:*
Substantial changes that affect the following aspects:
- Environmental control and monitoring
- Primary packaging
- Packaging validation

- Sterilisation process
- Sterilisation validation
- EtO residues
- Pyrogenicity
- Sterility during the shelf-life of the product

*Medical device, class I with measurement function*

Substantial changes can affect accuracy

- Calibration and calibration interval

*Medical device, class IIb und III*

- Substantial changes of the quality system
- Significant changes of the product
- New products outside of the certified product range

*Medical device, class IIa und IIb*

- Substantial changes of the quality system
- New product families outside of the certified product range

- Products using new technologies
- Products that are clinically applied in another way than the certified products

*... because Annex II CE certificates refer to specified, limited product ranges or to a special technology or a specified clinical application.*

*Medical device, class III*

- Substantial changes of the quality system
- Significant changes of the product
- New products outside of the certified product range

## 21.9. Approach of Implementing Modifications on CE-marked Medical Devices

The first step is the risk assessment of the planned modifications, followed by the assessment whether or not this modification has to be reported or even pre-authorised. Then, the technical documentation of the affected medical device has to be updated.

The following parts usually have to be updated:

- Risk management file
- Clinical assessment
- List of the essential requirements
- List of the applied standards
- Labelling

This is followed by an update of the quality manual or of the affected standard operating procedures.

### 21.10. Risk Assessment of Modifications

In general two factors are assessed:

- The probability of occurrence of an incident *and*
- The severity of the consequence after the incident has occurred

Moreover, it is differentiated between ...

- Non-tolerable risks (from a certain level the risk outweighs the benefit. These risks have to be avoided).)
- Tolerable risks (benefit outweighs the risk) *and*
- Negligible risks

### 21.11. Summary

Products are considered as safe if they a free from non-tolerable risks. Risks have to be reduced using state-of-the-art and preventable risks must be avoided. When talking about risks – this means the risk to patients but also to users or third parties.

The manufacturer has to perform a continuous process of risk reduction taking into account all facts and possible controls. All this has to be reflected in an update of the technical documentation.

21.12. Test Your Knowledge

| | |
|---|---|
| **Q1:** | List at least 4 potential triggers for the modification of a medical device. |
| **A1:** | - Requirements of the market |
| | - Modification of material |
| | - New standards |
| | - New approval requirements |
| | - Economic considerations |

| | |
|---|---|
| **Q2:** | What type of modifications can be differentiated? |
| **A2:** | - Product modifications |
| | - Changes in the quality system |
| | - Significant changes (product-related) |
| | - Non-significant changes (product-related) |
| | - Substantial changes (QS-related) |
| | - Non-substantial changes (QS-related) |

| | |
|---|---|
| **Q3:** | List three possible modifications of the quality system? |
| **A3:** | New product line, change in validation, sterilization, material, organisational structure, new suppliers … |

| | |
|---|---|
| **Q4:** | What changes have to be reported to the Notified Body? |
| **A4:** | Significant (product-related) and substantial (QS-related) changes. |

| | |
|---|---|
| **Q5:** | A manufacturer changes the calibration interval for a medical device with a measurement function. Does this change need to be authorised? |
| **A5:** | Yes. |

| | |
|---|---|
| **Q6:** | What aspects are important when assessing risks? |
| **A6:** | - Probability of occurrence of an incident |
| | - Severity of the consequences when incident occurred |

| Q7: | A medical device is considered as safe ... |
| --- | --- |
| A7: | ... if it is free of non-tolerable risks. |

21.13. References

- EPA BVMed, Berlin, Abbildung 22/2

- Menzl, Stefan, Regulatory Compliance and medical devices, FORUM Seminar 2011

- Scholtz, Sibylle, Vom Lesestein zur LASIK, Aachen 2006.

- Apple, David J., Sir Harold Ridley and his fight for sight, Slack, 2006

- www.rayner.com/history/1949

- Scholtz, Sibylle, An ophthalmic success story, The history of IOL materials, CRSTE 9/10 2006

- Scientific poster at ESCRS congress 2006, Scholtz, Sibylle / Weber, Klaus: From London around he World – Sir Harold Ridley and his idea of Intraocular Lenses conquered the world

- Auffarth, Gerd U., Accommodating IOLs, IOL-Info 2004 – 2005, Cologne 2004

- NB-MED/2.5.2/Rec2 Reporting of design changes and changes of the quality system

- MDD 93/42/EEC (http://ec.europa.eu/health/medical-devices/files/revision-docs/2007-47-en_en.pdf (English) und http://eur-lex.europa.eu/LexUriServ/LexUriServ.do?uri=OJ:L:2007:247:0021:0055:de:PDF (German))

- www.tecnisiol.com/eu/physician.htm

- www.amo-inc.com/products/cataract

- www.tecnismultifocal.com/us/full-range-of-vision.php

**Chapter 22: MDD Revision**

*Dr. Stefan Menzl*

22.1. Learning Objective

All three European medical device directives (IVD, AIMD, MDD) are currently under revision and will be modified to a certain extent. This chapter will shortly describe the planned modifications and discuss potential consequences.

22.1. Introduction

The European regulatory framework for medical devices has proven its value during the last 20 years. It allowed innovations in the field of medical technology to be implemented on the healthcare market relatively quickly. New technologies are available in the EU about 2 years ahead of the US and about 5 years ahead of Japan.

Nevertheless, a draft of a revised legislation was published on September 12, 2012 – after several delays.

22.2. What will be New? What will Prevail? ... and Why?

The revised legislation is based on the strengths of existing regulations. That means, it is an integrated approach to ensure

- The protection of patients, users and third parties
- That innovations can quickly enter the market
- The competition between the companies
- The principles of the free movement of goods within Europe.

In the nearer future, medical devices will be regulated by regulations rather than by directives in Europe. This means that there will no longer be the need to transpose the European text into the national laws as regulations, different to directives, are directly applicable. This will lead to uniform requirements in Europe.

The new requirements will be legally binding directly after adoption by the European Parliament and after a fixed transition period of probably 3 years.

Existing requirements were continuously enhanced and further defined in the new medical device regulation. It is planned to transfer the MDD, the directive on active

259

implantable devices as well as the regulations concerning material of animal origin into one regulation only.

Invasive products that do not have a medical intended use (this relates to several esthetical implants) are likely also be regulated under the revised medical device regulation.

The re-use of medical devices that are defined by the manufacturer as single-use products will be highly regulated. Moreover, so-called "in house" tests, genetic tests and medical devices used for diagnostic purposes from a distance will also be affected by the EU regulation.

Another focus of the future medical device regulation will be on the accreditation of the Notified Bodies and on the surveillance of their activities. The objective is to define and implement uniform standards for the conformity assessment carried out by the Notified Bodies.

As far as the clinical assessment of safety and efficacy is concerned, the new regulation is expected to result in an increase of clinical trials. Even after placing new medical devices on the market, manufacturer will be obliged to perform in defined intervals so-called „post market clinical follow up". One source of input to this follow up can be post-market clinical trials. It is expected that processes how to initiate and conduct clinical trials will be standardized. Moreover, a more sophisticated coordination of safety-related aspects of the assessment of multicenter and multinational trials will be established.

As already mentioned, a focus of the regulation will be on the definition of measures applicable after new medical devices have been placed on the market, e.g. a timely response to safety deficiencies raised by the national competent surveillance authority.

The regulation is expected to aim at a better coordination of the analysis of vigilance cases and new information resulting from market surveillance. This will lead to a higher level of interaction with experts from the healthcare sector as well as patients.

The task of the competent national authorities in the future will be

- To drive the harmonisation of regulations
- To approve clinical trials
- To accredit/notify Notified Bodies
- To assess vigilance cases
- To conduct market surveillance
- To implement measures in the companies and in the market

High priority is given to transparency. In order to strengthen transparency, the implementation of a central database will take place. This database will list all medical devices available on the European market as well as the responsible economic players in the marketplace. Data relating to product safety and effectiveness of medical devices will be published and the traceability of medical devices will be improved by the so-called „unique device identifier".

Considering the high frequency of companies placing product innovations on the market as well as of continuous development of existing technologies it does make sense that the competent authorities look for support by scientific expert panels when assessing these technologies.

The EU Commission relies on a network of reference laboratories and on already existing scientific institutions within the EU. The Commission will establish a so-called "Medical Device Coordination Group" which will likely take part in the conformity assessment procedure of implants, class III products and innovative technologies.

## 22.3. What Kind of Influence will the New Regulation have on the Notified Bodies?

First of all, the Notified Bodies will have to "face" the so-called "Medical Device Coordination Group", chaired by the EU Commission.

An assessment of the activities of Notified Bodies by other Notified Bodies or by appointed specialists could become a general rule. Moreover, audits by several representatives of different Notified Bodies or state authorities could be defined. Furthermore, unannounced audits will be carried out. This might be a logistical challenge for both – the Notified Body and the manufacturer because relevant documents and experts have to be provided at short notice.

As far as the Notified Bodies are concerned, it is expected that the qualifications of the auditors as well as the assessors for technical documentation will become more sophisticated.

Article 44 of the draft of the new medical device regulation will be of particular importance. As already mentioned, the EU Commission will likely establish the Medical Device Coordination Group (MDCG). Once this group is established, the Notified Bodies are asked to inform the Commission and the MDCG upon receiving a request for a conformity assessment of new class III or other 'high risk' medical devices. The Notified Body then has to provide the draft of the instructions for use as well as a summary report (relating to the safety and clinical effectiveness of the product).

The MDCG can ask – within the period of 28 days – the Notified Body to provide a summary report of the conformity assessment which will be further examined by the MDCG. Nevertheless, the MDCG has to justify their request. The Notified Body will inform the manufacturer within 5 days of receiving the request by the MDCG.

In the following 60 days, the MDCG can address questions or comments concerning the conformity assessment to the Notified Body. It is planned that the MDCG has the right to ask for further information and data within a 30-day period. This may include receiving a product type or even a visit at the manufacturer's site. In case the manufacturer or the Notified Body did not provide requested data, the 60-day assessment procedure will be suspended and a so called 'stop clock' will occur (which ultimately leads to a prolonged overall assessment time). Finally, the MDCG will give a recommendation to the Notified Body that has to take this recommendation into account when issuing a CE certificate or final conformity assessment report. Therefore, it is expected that no Notified Body will issue a CE certificate after having received a negative reply by the MDCG. This procedure may also be extended to other medical devices (beyond class III products) but only for a (not yet defined) certain period of time.

Criteria for this could be:

- Uniqueness of a product or of the applied technology
- Additional risks triggered by new components, material or impact in case of default
- An increased rate of serious adverse events for a defined product category

- A significant difference in the conformity assessment of a technology by different Notified Bodies
- On suspicion of a public health risk by a specific product category

## 22.4. Summary

Time will tell which of the described aspects will be incorporated in the final medical device regulation and how the implementation will be handled in detail. Without any doubt, the requirements to be met by the manufacturer and the Notified Body will get more stringent.

In the interest of patients, physicians, manufacturers of medical devices, payers and authority representatives, the new EU regulation should represent a reliable regulatory framework in order to further harmonize and reduce the fragmentation of national specificities.

A European database, a uniform definition for the implementation of clinical trials, the market accessibility for medical devices and a standardized procedure for reporting and assessing of adverse events would contribute to this.

## 22.5. References

- http://ec.europa.eu/health/medical-devices/documents/revision/index_en.htm
- www.bvmed.de/stepone/data/downloads/7b/e4/00/bvmed-jahresbericht2012.pdf
- http://eur-lex.europa.eu/LexUriServ/LexUriServ.do?uri=COM:2012:0542:FIN:EN:PDF

MDD Revision

264

## Chapter 23: Final Case Study with Sample Solutions

*Dr. Stefan Menzl, Dr. Sibylle Scholtz, Dr. Carsten Rupprath, Myriam Becker*

23.1. Learning Objective

In this chapter you will apply the acquired knowledge (acquired by reading this book) on a real case from everyday practice.

23.2. Case Study: Approval of a Medical Device

We now want to give you an impression of the reality and complexity of the registration procedures of medical devices via a fictitious but practice-oriented example. Any resemblance to real conditions and people are purely coincidental and is not intended.

23.3. Preliminary Remarks

In this particular case we ask you to collect input for an approval plan. The focus is on real-life relevance. The basic aim of this case task is the registration of a medical device by complying with the interlinked laws and regulations and the deducted requirements. Moreover, the primacy of economics, result relevance and earnings relevance has to be taken into consideration.

All aspects that you have learnt while reading this book will come together to a big picture. In the end you will have put together a registration strategy to which you will have substantially contributed by answering relevant questions.

Some questions will require you to rely on your creative intelligence and to gather additional knowledge by literature or web research. This approach does represent a scientifically proven way of gathering information that is frequently applied in professional life.

It is a demanding starting point, but this task will also bring to you new findings. Please use the web in order to get up-to-date information or guidance.

We wish you all the best!

23.4. The Project

This registration management task case has an interdisciplinary character. In the everyday reality of the healthcare industry this is rather the rule than the exception.

Experts from different departments come together in a project in order to work at a high level on the solution of a specific problem or task.

## Business Plan of the Innovative Optics AG (INO-AG)

You are a member of a team of external consultants of the renowned, worldwide operating consultancy company McBright. This company was engaged by the INO-AG to get new medical devices registered on the basis of defined parameters and business plan details. These registrations are meant to secure the future of the INO-AG. Your supervisor expects you to contribute to the registration plan by answering the questions (at the end of this document).

The consultancy company you are working for has to develop a phased registration and launch plan with timelines for the medical devices taking into consideration key aspects and criteria. It is not only about details or summarizing isolated facts but to discuss the overall situation with all possible options.

CAVE!

Please also describe needed capacities (staff, R&D, financials, ...) for the company when answering the questions. Please give as precise and as detailed objective reasons as possible to support your answers and suggestions.

The strategic registration plan should consist of the following elements:

- Definition and classification of the product
- Target countries and national requirements

*For EU countries*

- Registration details (i.e. planned production numbers, future product development, best strategic market entry)
- Order of the registration in the chosen countries
- Norms to be applied
- Harmonized norms
- Clinical assessment and clinical studies
- Labelling
- Requirements to the quality system
- Free sales certificates

- Provision of the product/registration in other countries
- Necessary post-market surveillance processes

Background information
- Business Plan of the INO-AG

The following business plan and its content are reality-oriented. The defined product is registered in many countries all over the world. In order to ensure data protection, this case task has been made anonymous.

# Business Plan of the Innovative Optics AG (INO-AG)

*Registration and launch of IOLs and implantation systems as medical devices*

<u>By</u>

Prof. Dr. Anonymous

Dr. Secret

<u>Date</u>

August 2013

<u>Preface</u>

INO-AG has focussed from the beginning on ophthalmologic products. It develops, manufactures and sells contact lens care products and products for ophthalmic surgery. All in all, the company has three main areas of expertise:

- Contact lens care
- Product for the treatment of cataract
- Refractive products (laser technology)

Refractive products are mainly used in ophthalmic surgery, for example laser machines and other medical-technical devices to measure people´s vision. The Bavarian company, located in Nonamecity, belongs to the worldwide operating INNOVA Group (New York). There are subsidiaries in various countries, on all continents. The European headquarter is in Bavaria which is responsible for Europe, Africa and Asian-Pacific area.

In cooperation with another company (company W. Unknown) innovative IOLs were developed in order to treat cataract. These IOLs are sold in a so-called pre-loaded configuration. This significantly improves the usage comfort and safety during the operation. The products are now ready to go into production and will be manufactured after registration in the US and Europe and will be sold globally. Results from initial clinical studies were positive and showed excellent results even in difficult indications. Therefore there is strong demand by ophthalmic surgeons who want to implant these IOLs.

The INO-AG wants to get first of all the CE certificate for the product and via this certificate the registration as a medical device. As far as quality assurance is concerned ISO 13485 as well as DIN ISO 9000 have to be taken into consideration.

In order to ensure a quick market entry, INO-AG asks a renowned external consultancy company to take care of the registration. Placing these products on the market will be a key success factor for INO-AG to secure its future.

The promotion of the registered products in further global markets will be the key activity of the INO-AG in the next years.

Facts

INO-AG is manufacturer and distributing company. The shareholders have invested over the last years significant intellectual, financial and time resources in order to get the product portfolio which is now promoted nationally and internationally by the INO-AG.

The activities are based on the existing sales divisions, research and design results as well as manufacturing partners of the INO-AG.

The focus – as far as selling and registration is concerned – will be in the first line on Germany. But existing international networks with potential customers and cooperation partners will be key for the future promotion of the medical device and for the international growth and the company's success. In a second phase, new markets in Europe, the Middle East and Africa will be targeted. In a third phase, the promotion of the products will be extended to other international markets.

Market

The INO-AG wants to be market leader in a few central markets with the defined product portfolio and its costs structure in order to be able to determine the prices and to be highly profitable. This is especially true for the market segments "refractive surgery" and "cataract surgery".

The demand for technically advanced, cost-efficient and easy to handle medical devices does exist – especially when considering the cost pressure in the healthcare sector. The implantation of easy-to-handle IOLs in a pre-loaded version (which will require only little training of surgeon and surgical nurses to be handled correctly) has a high market potential worldwide. Moreover, this segment of the economy can benefit from an ageing population and the age pyramid especially in highly developed countries.

But the registration of this innovative medical device also is a big step in the fight against vision impairment in elderly people. This technology will help to improve the quality of life and will maintain the autonomy of an ageing population worldwide. This technology also has socio-political dimensions that should not be underestimated.

Cataract surgery is the No1 surgery worldwide. The easy handling is especially important for surgeons in developing countries who have to carry out surgeries under difficult hygienic conditions and who lack first-class training but want to improve the vision of their patients.

SUMMARY

Parameters

The INO-AG is a young company that is built on European certificates and registrations in the area of medical devices. The company aims at the worldwide promotion of innovative products as well as at the market leadership in defined target countries.

In 2013, the promotion of this innovative IOL in a pre-loaded version is planned to start in defined European target countries. In the following years, further countries in Europe, the Middle East, and Africa will be targeted.

The company needs 1.000.000 € within two years (350.000 € in 2012 and 650.000 € in 2013) mainly for registration purposes. The main investments for the product promotion therefore will be planned for the following financial year.

Earnings before taxes are as follows:

- 2010:   -415.000 €
- 2011:   +612.000 €
- 2012:   +1.675.000 €
- 2013:   +4.030.000 €

Investments for the promotion and launch of the medical device will be funded by the cash flow of the company.

The trademarks and the web domain have been secured. The website of the INO-AG has been launched in German and English. Further languages (French, Spanish and Arabic) are underway.

The innovative IOL in the pre-loaded version has been positively tested in clinical studies. The first serial production has started without malfunctions. Promotional and scientific material has been prepared. Modifications of the material for further market segments still have to take place.


CORPORATE PHILOSOPHY

Mission statement and company´s objectives:

The key objective of the INO-AG is to sell medical devices nationally and internationally in specific indications that need innovative solutions/products. The company focuses on disorders where economically advanced treatments are not available. The products of the INO-AG are innovative, easy to apply and technically advanced.

The corporate philosophy reflects the experience of the company's founders (who are also ophthalmic surgeons) who worked as doctors in the US, Western Europe, Africa, Asia and the Middle East.

It has been proven that products that are easy to apply significantly increase the safety in the operating theatre as well as shorten the operation time which results in a cost reduction. Not only in developing countries but also in highly developed countries is the pressure to reduce costs and to rationalise a challenge for MD manufacturers, hospital managements, doctors and health insurances. This led to the development of ideas and concepts that take this trend into consideration and to design new strategies and technologies.


BUSINESS MODEL

Due to the high investments in research and development in the health care sector and due to the capital and resource commitment, the INO-AG actively promotes interdisciplinary interlinking of knowledge, procedures and existing products in order to create new, innovative concepts. By cooperating with the scientific community and the industry, product innovations are being promoted that are more innovative and less expensive than the products promoted by the competition. The company uses state-of-the-art knowledge management tools as well as a number of cooperation partners and network partners.

A flexible organisation of permanent employees and freelancers with different knowledge and expertise was set up. This workforce is the owner of the innovation expertise. The INO-AG takes care of the interface management, the knowledge management, the transfer process control as well as R&D. The INO-AG owns the R&D competencies, trademarks, patents and product rights, as well as marketing competencies and licences. The tasks of a sales company are taken over by the INO-AG in countries where subsidiaries already exist. In countries without own subsidiaries, the INO-AG cooperates with distributors that may also take care of the registration. Owner of the registration is the INO-AG wherever this is possible.

Legal facts

The INO-AG was founded in 2002 with a nominal capital of 25.000 €.

Shareholder: Max J. Example

Further shareholders will be taken on.

Headquarter of the company is in Nonamecity /Germany.

The INO-AG is an independent company that designs, manufactures and distributes medical devices.

The business development will be on an international scope. The way that foreign business will be exercised (via company participations or via licence agreements) has to be determined by the (new) shareholder meeting of the INO-AG. The registration strategy for the international promotion of the product has been outsourced to the renowned Clinical Research Organization (CRO).

Product status

All products are ready to be sold and have been designed according to EU regulations or are about to be registered. The INO-AG is already planning to enter the markets in Europe, US, China, Israel, Palestine, Turkey, Egypt, Jordan, Syria, Iraq, Iran, Scandinavia, India, Sri Lanka and Romania (partly with strategic alliance and cooperation partners).

Already underway are treaties with partners in Austria, Switzerland, Sudan, Balkan States, Estonia, Latvia and Lithuania. In the meantime, first orders were received from Africa (Swaziland and Zimbabwe). Market authorization for some of these markets can be supported by CE certificates.

Prerequisite for the orders is – due to the information by the partners – the registration in the EU.

Products and trademarks

The following product will be distributed (it is already registered, the trademark is also registered): preloaded delivery system containing 1-piece soft acrylic intraocular lens, Vision 1-piece lens as model VCB00

Markets

The INO-AG wants to be market leader in some central markets with its defined product portfolio and cost structures in order to be highly profitable. This is especially true for the market segments "refractive surgery" and "cataract surgery".

The demand for technically advanced, cost-efficient and easy to handle medical devices does exist – especially when considering the cost pressure in the healthcare sector. The implantation of easy-to-handle IOLs in a pre-loaded version (which will require only few trainings by surgeon and surgical nurses to be handled correctly) has a high market potential worldwide. Moreover, this segment of the economy can benefit from an ageing population and the age pyramid especially in highly developed countries.

As already mentioned, cataract surgery is the No1 surgery worldwide. The easy handling is especially important for surgeons in developing countries who have to carry out operations under difficult hygienic conditions and who lack first-class training but want to improve the vision of their patients.

Target countries / phase 1: The focus concerning registration and distribution is on Germany. Nevertheless, the future distribution strategy of the medical device (due to existing international networks to potential customers and cooperation partners) is key for the international growth and success of the company.

Target countries / phase 2 & 3: In a second step, markets in Europe, the Middle East and Africa are to be targeted. In a third step, the product portfolio will be promoted to other international markets.

Strategic partners

The founders of the INO-AG already work closely together for some time with a number of strategic partners. The following list only includes strategic partnerships in Sales & Marketing:

| Name of the partner | Definition of partnership | Contract status |
|---|---|---|
| Dr. Paul Wohlgemuth, Inthemiddleofnowhere, Germany | Microbiological analytics | Silent partner |
| Dr. A. Bohra Ltd, Bangalore, India | Distribution India / Sri Lanka | In preparation |
| Dr. U. Gumusbas, Ankara, Turkey | Distribution Turkey | In preparation |
| Prof. R. de Risio, Sacramento, USA | Regulatory Affairs Consultant | Silent partner |
| Dr. A. Najm, Isfahan, Iran | Iran | Contract signed (incl. delivery), registration for Iran ongoing |
| Dr. F. Sangak, Cairo, Egypt | Egypt | Letter of intent |
| Dr. M. Saleem, Amman , Jordan | Distribution Arabic/Gulf region | Letter of intent |
| Dr. I. al-Haitam, Basra, Iraq | Iraq | Letter of intent |
| Dr. S. Malki, Aleppo, Syria | Syria | Letter of intent |
| Dr. A. Abusharekh, Jericho, Palestine | Palestine | Contract signed (incl. delivery) |
| Dr. A. Scandinavius, Oslo, Sweden | Distribution WTAGmed and WTAGvet | In preparation |

Technology

Over time – due to a natural ageing process – the natural lens of the eye becomes opaque which results in significant visual impairment. In cataract surgery the lens is fragmented via ultrasonic technology, removed from the eye and replaced by an artificial lens (implant). This artificial lens (IOL) can be made from different material (PMMA, silicone, hydrophobic or hydrophilic acrylate, …), can have different optical quality (asphericity, acromatic lenses …) and can be a monofocal or multifocal lens.

Monofocal lenses have ONE focal point and enable a good far prescription (reading glasses have to be worn for reading). Multifocal lenses have SEVERAL focal points and work like varifocal glasses. The patient has clear vision in near and far distances (he usually does not need reading glasses). INO-AG has ongoing research concerning monofocal and multifocal lenses that will be placed on the market in so-called "pre-loaded systems".

This IOL will be implanted into the eye via an implantation system that works like a syringe. This IOL will be placed in the implantation system together with a lubrication agent (viscoelastic).

Classical technology asks for a complicated procedure: The surgical nurse or the ophthalmic surgeon has to take the implantation system from its sterile packaging, to fill it with the viscoelastic. Then he has to remove the IOL from its sterile packaging and to load it into the implantation system. Only then can he implant the lens into the eye. The innovative concept of the INO-AG has a fully configured system – the lens and the viscoelastic is already loaded into the implantation system and put into a single sterile packaging. This single-step procedure shortens operation time, sterility is always ensured and even less experienced surgeons and surgical nurses can easily use this technology.

Planned developments

The necessary distribution planning has already started. Studies comparing the efficacy and efficiency of the product are needed. The results could also be derived from the experience of the users but reliable scientific data has to be generated.

It is assumed that the preloaded configuration will offer clear benefits as far as safety, efficacy and efficiency is concerned. The company will apply for a FDA registration in due time – depending on the results of the EU registration.

The staff size of permanent employees will remain 25 in 2012 and will be increased to 60 in the following years. The senior expert pool of 16 freelancers will be increased due to the success of the business development.

## Registration alternatives

The newly designed preloaded IOL consists of several components (lens, viscoelastic, cartridge, implantation system). Every component is also a medical device. First of all it should be explored which registration alternatives do exist and which of these alternatives fits best to the company's strategy.

### 23.5. Case Task

Jot down how a registration of the single components could look like and what would be the pros and cons of this strategy in your opinion.

Then decide whether or not there is an opportunity to register the product as a whole and what would be the pros and cons of this strategy.

Please take into consideration that the company is planning to place further lenses in a preloaded system on the market.

What consequences will the chosen strategy for the CE certification have on the registration in other target countries?

## The business environment

"The healthcare market is an international, promising and also ethically demanding business with excellent growth opportunities." (Source: Apollon Hochschule der Gesundheitswirtschaft).

As far as the vast expansion potentials of the pre-loaded version is concerned because of growing number of surgeries (cataract and refractive), it can be assumed that the INO-AG will launch its product in a growth market. Moreover, it is the perfect time for a launch. Due to the strained financial situation in the healthcare sector, the demand for the products of the INO-AG – if promoted the right way – will be very high. This is especially true because of the excellent cost effectiveness.

## The healthcare market

As shown on the last pages, the market is very big and diversified, that it makes sense to concentrate on core areas. In the next paragraphs only the opportunities for

Germany will be described. Nevertheless, the global market seems to be even more interesting for the INO-AG.

Status quo Germany

The healthcare market in Germany has reached a high standard – if compared historically and internationally. This standard (or to be precise: the use of new treatment regimen) is somewhat threatened by the growth in expenditure and consequently the impact on the contribution rates of the health insurances. Only few countries in the world have a comparable social security system. Nevertheless, any cost savings in this area is of high importance: approx. 20% of the working population in Germany is employed in the healthcare system. About 150.000 people work for approx. 260 statutory health insurances. 2.200 hospitals ensure a high-quality in-patient hospital care. 311.000 doctors are working in this sector. The total budget of the German healthcare system amounted to 285 billion € in 2006. Since 1970, the percentage of the healthcare cost increased from 6.3% to 10.7% of the entire national economy.

The health is perceived to be the most important asset and therefore there is high willingness to pay for healthcare services. But statutorily insured patients often consider the healthcare services provided as "free of charge" and consequently "consume" these services in total ignorance. This limits the willingness to pay for certain healthcare services. On the other hand, the willingness to pay for these services increases disproportionately with the growing wealth and increased awareness of the population. The needed budget comes from the rising incomes. The Germans have got used to a high standard.

The age structure of the German population has changed drastically since the 1980s. The number of people below 21 years of age is as low as 21.5%. More than one-fifth of the population is over 60 years of age. Over the next centuries, this trend will be aggravated. Moreover, life expectancy is rising. This will result in less revenues and higher expenditures in health & welfare.

When transforming the healthcare system, the objective has to be to maintain the service level while implementing regulating and controlling mechanisms. Treatment efficiency and control of treatment costs are increasingly in the focus of interest. The responsibility of the individual has to be reinforced. All this ensures that this is an excellent time to launch the products of the INO-AG.

Competitive pressure is increased by advanced technology that is commonly used in the highly developed industrial countries. The design of new innovative products, the transfer of research results into practical applications takes place at increasingly short intervals. This increases the pressure to adapt in the market. The effects in the labour markets are a signal of this development that cannot be overlooked.

*"Progress in every age results only from the fact that there are some men and women who refuse to believe that what they know to be right cannot be done."*
Russell W. Davenport, 1899 – 1954

A lot of economic experts believe that the healthcare market will show high growth rates at least until 2020 and even beyond. In the medium term, nearly no other field will have the potential for additional jobs – most economists are sure of this. It is assumed that until 2020 there will be approx. 1 million additional jobs. Other experts say that healthcare services are the "basic innovation of the 21st century" and talk about the "mega-market healthcare". The information era is coming to an end. The desire for a long, carefree life will trigger the next growth wave of the global economy. The bigger growth will take place in the segments that are not exclusively financed by the statutory health insurances.

Competition

The competitive structure is characterized by big multinational companies. All big pharmaceutical companies and MedTech companies have products on the market that will be partly affected by the products of the INO-AG. But as the innovative products of the INO-AG offer new answers to new challenges, the INO-AG does not have to be afraid of the competition. Moreover, the demand for this kind of products is very high throughout the world that even if the INO-AG only had a share of 0.01% of the German market, it would have sales of at least 50 million €. In addition, there is no company on the market at the moment that has a comparable medical or technical focus. Moreover, a cost comparison shows that the price of the majority of the INO-AG is way below the price of the competition. So, there is no risk for the marketability of the products. The USPs (unique selling points) of the INO-AG products in a broad indication spectrum cannot be offered by the competition in the foreseeable future.

Prerequisites for success

The prerequisites for success in the healthcare market result from the quality and the cost/benefit-ratio of the products of the INO-AG as well as from its Marketing strength and sustained funding. Moreover, there are the following prerequisites of success:

- A clear and easily understandable company profile
- An experienced management team with high-quality network relations
- Cooperation and network partners that are established on the market
- Good network within the scientific development areas
- A highly professional communication strategy of the company
- An existing high-quality and innovative product portfolio
- Already tested market niches with high potential
- Solid financial and development plan
- A highly decentralized and split main target market and niche market
- USPs in some indication areas
- Offers with high benefit-cost ratio

Risks

For a young, fast growing organization the following risks have to be taken primarily into consideration:

- Strong competitors
- Competitors copying the products
- In case the company does not succeed in getting the products registered
- Too much objectives (building the organization and growing the products) at the same time
- Too long start-up time which will significantly reduce the strategic and innovative leadership
- Especially during the growth period: acquisition of unsuitable human resources
- Insufficient strategic communication concept of the company
- Insufficient IT preconditions
- Inadequate acquisition performance and project management resources
- Potential quality problems in case of continuous operation of the plants
- Low capital endowment

## SPECIFIC TASK

### Registration strategy

As already mentioned, the newly developed IOL in pre-loaded configuration consists of several components (lens, viscoelastic, cartridge, implantation system). Please describe what different kinds of registration possibilities there are and which of these possibilities suits best taking the company's strategy into consideration.

Describe shortly how the registration of the single components could look like and what pros and cons you see in this strategy. Then please decide whether or not you see the possibility to register the product as a whole and what would be the pros and cons of this strategy. Please also consider that the company is planning to put further lenses in a pre-loaded configuration on the market shortly. What consequences will the chosen strategy have on the CE certification and registration in the already mentioned additional target markets?

### Order of the regional and national registrations

The company has a short-term, a medium-term and a long-term strategy which also reflects in the order of the registration plan. Please work on a proposal of the order of the regional and national registrations which reflects the company's overall strategy. Please consider the time to achieve a specific national registration, the documentation in place at the time of application, effort and costs. Also, reimbursement aspects can be taken into consideration. Please explain your proposal extensively.

### Applicable norms

Prepare a list of applicable norms for the relevant product components or products. Also consider European harmonized standards.

### Applicable legal requirements

Prepare a list of the most important directives. Also name regional or national applicable legal requirements in order to register the products.

Classification of the product/the products

Classify the given product/products (medical device, in-vitro diagnostics, medicinal products, cosmetic, …). Take into account the corresponding directive. List the classification regulations that you applied as well as the assumptions you made.

Technical documentation

Checklist of the essential requirements. Define the basic framework of the checklist of the essential requirements.

Risk analysis

Describe which departments have to be involved for preparing the risk analysis for the given product/products and why. What special risks would you address?

Chemical, physical and biological tests

Describe, whether or not (and which) preclinical studies have to be conducted for the given product/products and if YES – why? Do you think that biocompatibility tests are necessary for the given product/products? Under which conditions would it be possible to refrain from these biocompatibility tests?

Microbiological safety, Material of animal origin

Describe which aspects are covered in this chapter. The viscoelastic that improves the mobility of the implantation system and fosters the safe implantation of the IOL into the eye, is a hyaluronic acid product. R&D can purchase this hyaluronic acid from two different suppliers. The first supplier produces hyaluronic acid by microbial fermentation. The second supplier produces hyaluronic acid from raw material that is made from rooster combs. Assess both alternatives as far as safety, classification of the products, long-term reliability of the supplier, extent of the required technical documentation for the registration, the registration process, effects on the registration in other countries (outside of the EU), required manufacturing licences. Please give a clear recommendation for the management. Keep in mind that viscoelastics are classified in some markets as medicinal product.

## Coated medical devices

Is this chapter of the technical documentation relevant for the given product/products? If YES, which aspects should be covered in this chapter?

## Clinical data

The company manufactures IOLs already for centuries and therefore has extensive clinical data. Compared to the CE-certified previous models, the new IOL does not differ in indication, material or design. What's new is the combination of the IOL, viscoelastic and implantation system. Is – from your point of view – a new clinical study necessary? If YES: which aspects have to be covered by the study? Does a clinical assessment have to be conducted or can the company fall back on the updated clinical assessment of the previous IOL? Please explain your recommendation extensively.

## Qualification of the packaging and shelf life

Which tests should be carried out and documented?

The Marketing department asked for extensive testing while the product is already dispatched in its packaging to various target countries. The integrity of the packaging and the product will be assessed in the target countries.

The Sales Manager who is responsible for Saudi Arabia even asks that the product is stored at a temperature of over 40° Celsius for two weeks and then to assess the integrity of the product and product quality. He argues that the products sometimes will stay in customs at Abu Dhabi before being transferred to the sales organization.

When the Sales Manager for Spain and Portugal hears this, he asks the product to be stored for 3 – 4 weeks at 60° Celsius because the products are transported by truck to Spain and Portugal. This transport can last for 3 to 4 weeks, and in summer it could easily be about 60° Celsius.

The Head of R&D argues that he has performed these tests already with the similar previous model. He will provide a written justification that these tests can be also used for the new product. One hundred thousand Euro could be saved by this and the product could be placed on the market much earlier.

Assess the different points of view and give a recommendation for the management how to proceed.

## Product labelling

List the mandatory information that has to be printed on the product labelling.

The Product Manager does not want a big label with all mandatory information because this would disrupt the looks of the packaging. He wants this information to be printed on several small labels. What do you think about this?

The Sales Manager for Spain and Portugal wants the name of the product (that defines the kind of the product) to be translated into Spanish and Portuguese. The head of production opposes this and mentions that all mandatory information is covered by EN 980 symbols and therefore translations are not necessary. Do you agree?

## Promotional material

The main European congress in this field will take place in autumn. It is absolutely essential for the company to present and to promote the product at this congress. At the moment it looks like the registration approval will be received several weeks after the congress. The Sales Manager as well as the Head of Marketing want to actively promote the product at this congress with posters and promotional videos which show the product being applied. Moreover, Marketing wants to give samples of the product to potential customers. The Head of Clinical Research has agreed to have a live surgery (where the product is used) from the operation theatre broadcasted to the congress.

Assess these proposals taking into account the registration approval is not likely to be received at the time of the congress. Please give a recommendation.

Is it allowed to present not registered products at a congress? If YES: what are the prerequisites? Is it allowed to give samples to potential customers? Is it possible to give out non-sterile products that are labeled accordingly? What about the broadcast from the operation theatre? Could you think of a scenario where this would be possible?

## Sterilization

Which documents on sterilization are usually part of the technical documentation?

## Final evaluation

The risk of the product and of its application has been reduced by various iterative processes. Nevertheless, the final evaluation shows a residual risk. Is it possible to have the product registered nevertheless? What are potential prerequisites? What has to be the conclusion of the final evaluation?

## Declaration of Conformity by the manufacturer

Draw up a draft of the Declaration of Conformity for the given product/products.

## 23.6. Sample Solution

### Author

Max Weber

### Summary

Input for a registration plan was required based on a defined, fictitious business plan and expert reports of a company by working on given questions.

You are a member of a team of external consultants of the renowned, worldwide operating consultancy company McBright. This company was engaged by the INO-AG to get new medical devices registered on the basis of defined parameters and business plan details. These registrations are meant to secure the future of the INO-AG. Your supervisor expects you to contribute to the registration plan by answering the questions (at the end of this document).

It is important for the assessment of your solutions that you clearly explain your decisions and recommendations. In many cases there is not only ONE correct recommendation. In this case, it is key that you can list and explain the risks and opportunities of each scenario. Finally, you have to give a recommendation that is strategically feasible for the company.

## SPECIFIC TASK

---

**Q: Registration strategy**

As already mentioned, the newly developed IOL system in pre-loaded configuration consists of several components (lens, viscoelastic, cartridge,

---

284

implantation system). Please describe what different kinds of registration possibilities there are and which of these possibilities suits best taking the company's strategy into consideration.

Describe shortly how the registration of the single components could look like and what pros and cons you see in this strategy.

**A:**

Advantage of a single component registration:

This allows for maximum flexibility for future product modifications and modification of single components. Moreover, single components can be exchanged or replaced in case that the registration of one component is difficult in some markets (i.e. classification of the viscoelastic as medicinal product). To choose a viscoelastic that is not classified in a defined market as medicinal product can result in major cost savings and time savings (up to 5 years).

Disadvantage of a single component registration:

The disadvantage could be that after the CE registration every component may have to be registered in other target countries. Even small modifications in this case might be subject to notification obligation and authorization. This leads to additional work and expenses concerning the registration process and obtaining the authorizations.

For target countries outside the EU, so-called free sale certificates (FSCs) have to be obtained for every single component. This might lead to problems or limitations concerning the registration by countries that require FSCs from the manufacturing country (a registration of every component of the manufacturing country is the prerequisite for the issuing of an FSC and thus prerequisite for the authorization of a product by a country that requires a FSC from the manufacturing country).

Advantage of a system registration:

If all components are defined as ONE medical device, this product can get a CE-certificate. In this case, the single components do not have a CE mark but only the labelling and packaging of the end product have a CE mark affixed.

The advantage of this strategy is, that only ONE certification application has to be made. This leads to cost savings. Minor modifications of the components or of their production can be dealt with in the course of the supplier or components control.

Therefore not all modifications are subject to notification obligation or authorization. A further advantage is that a certificate is issued for a clearly defined product. This makes obtaining registrations outside of Europe easier.

It may be feasible to manufacture the product in a country that receives the registration at an early point in time. Thus, a FSC can be issued early in time and a registration application started much earlier.

<u>Disadvantage of a system registration</u>

This registration is less flexible as far as the exchange of components is concerned. A change of components would be a significant change of the product that is subject to notification. Maybe the end product will be classified to a higher risk class (due to a single component). The product may also be classified as medical device with drug component.

In a worst case scenario, the whole product could be classified as medicinal product due to the viscoelastic that is classified as medicine in some countries. Both scenarios would result in additional expenses and even to years of delay until the registration is obtained.

<u>Summary</u>

Both scenarios have advantages but also contain risks. It is a question of what reasons are provided to follow the recommendation!

---

**Q: Order of the regional and national registrations**

The company has a short-term, a medium-term and a long-term strategy that also reflects in the order of the registration plan. Please work on a proposal of the order of the regional and national registrations which reflects the company's overall strategy. Please consider the time to a national registration, the documentation in place at the time of application, effort and costs. Also, reimbursement aspects can be taken into consideration. Please explain your proposal extensively.

**A:**

When comparing the registration system worldwide, the European system of the CE certifications for medical devices is a very efficient and easy to plan system that is open for innovations and product development. The required expenses are manageable. The tests and the technical documentation can be used as a basis for further worldwide registrations. Due to this and the fact that the company is located in Europe and is closely connected with the European markets, it makes sense to get CE certificates for the new technology in a first step.

The focus target market is Germany. Eventually further European markets as well as markets in the Middle East and Africa will be targeted. The CE certifications enable the access to 28 countries. As many countries accept CE certifications, it is quite easy to extend the market access for example to Switzerland and other countries. Moreover, CE certifications are an accepted basis to get a registration in many countries of the Middle East (i.e. Turkey, Saudi Arabia, ...) and in Africa (Egypt, Morocco, South Africa, ...).

The registration in Asian markets is made significantly easier when CE certificates and technical documentation are available. Also FSCs that are essential for many countries outside of Europe, can be easily achieved when CE certificates are available.

Recommendation concerning the order of registration:
- European CE certificate (28 countries)
- Switzerland
- Eastern Europe
- Middle East
- Africa

---

**Q: Applicable norms**

Prepare a list of applicable norms for the relevant product components or products. Also consider European harmonized standards.

Please see Annex 2

Applicable legal requirements

Prepare a list of the most important directives. Also name regional or national applicable legal requirements in order to register the products.

**A:**

- Directive 93/42/EEC of 14 June 1993 concerning medical devices

- Directive 2003/32/EEC of 23 April 2003 introducing detailed specifications as regards the requirements laid down in Council Directive 93/42/EEC with respect to medical devices manufactured utilising tissues of animal origin
- Directive 2001/20/EEC of the European Parliament and of the Council of 4 April 2001 on the approximation of the laws, regulations and administrative provisions of the Member States relating to the implementation of good clinical practice in the conduct of clinical trials on medicinal products for human use
- Directive 2000/70/EEC of the European Parliament and of the Council of 16 November 2000 amending Council Directive 93/42/EEC as regards medical devices incorporating stable derivates of human blood or human plasma

**Q: Classification of the product/the products**

Classify the given product/products (medical device, in-vitro diagnostics, medicinal product, cosmetic, ...). Take into account the corresponding directive. List the classification regulations that you applied as well as the assumptions you made.

**A:**

IOL, viscoelastic, cartridge and implantation system are classified in Europe as medical devices. According to Directive 93/42/EEC the single components are classified as follows:

IOL

Class IIb

Rule 8: All implantable devices and long-term surgically invasive devices are in class IIb unless ...

Viscoelastic

Assumption 1: Hyaluronic acid is derived from rooster combs > Class III

Rule 17: All devices manufactured utilizing animal tissues or derivatives rendered non-viable are class III except where such devices are intended to come into contact with intact skin only.

Assumption 2: Hyaluronic acid from fermentation without using material of animal origin > class IIa

Rule 6: All surgically invasive devices intended for transient use are in class IIa unless they are ...

Implantation system: sterile single-use cartridge

Rule 6: All surgically invasive devices intended for transient use are in -> class IIa unless they are ...


Classification of the whole product as medical device

Assumption 1: Hyaluronic acid is derived from rooster combs > Class III

Rule 17: All devices manufactured utilizing animal tissues or derivatives rendered non-viable are class III except where such devices are intended to come into contact with intact skin only.

Assumption 2: Hyaluronic acid from fermentation without using material of animal origin > class IIb

Rule 8: All implantable devices and long-term surgically invasive devices are in class IIb unless ...

The highest risk class of one component defines the class of the medical device.

---

**Q: Technical documentation**

Checklist of the essential requirements. Define the basic framework of the checklist of the essential requirements.


**A:**

Have a look at the Essential Requirements Checklist, Annex I,

---

**Q: Risk analysis**

Describe which departments have to be involved for preparing the risk analysis for the given product/products and why. What special risks would you address?


**A:**

"An expert is someone who knows some of the worst mistakes that can be made in his subject, and how to avoid them." (Werner Heisenberg).

A cross-functional team is always needed in order to have an efficient risk management. Moreover, a risk analysis should always be drawn up by a cross-functional team.

ISO 14971 defines the responsibility of the management. The management provides adequate resources, appoints qualified staff and defines principles for risk acceptance according to the state of the art as well as to harmonized norms. Moreover, reviews at regular intervals have to take place to assess the effectiveness of the measures. The norm cannot provide clear targets for risk acceptance because of the diversity of medical devices and their applications.

It is reasonable that the review of the risk management is embedded in the management review of the QM system.

The staff that takes care of the risk management has to have the needed expertise, competence and the experience. That means that they have to know their way around medical devices (i.e. know-how on material, on manufacturing techniques, on harmonized norms as well as on product applications). Moreover they need to know about risk analysis techniques as well as procedures that are part of the risk management. The qualification of the staff has to be proven upon request.

The risk analysis and the risk management should be performed, involving at least the following company functions:

- R&D
- Clinical
- Regulatory Affairs
- Quality assurance
- Marketing/Sales
- Production
- Logistics
- Users (ideally externals)

**Q: Chemical, physical and biological tests**

Describe, whether and not (and which) preclinical studies have to be conducted for the given product/products and if YES – why?

**A:**

As the product comes into contact with human tissue and as the IOL remains implanted long-term, it is key to perform all relevant tests on biocompatibility, cytotoxicity and potential allergic reactions. This is also true for the viscoelastic, as it cannot be ruled out completely that parts of it remain in the eye after the operation.

Moreover, so-called "handability tests" should be performed in order to prove that the product can be used easily and safely in the everyday OR environment.

**Q: Biocompatibility tests**

Do you think that biocompatibility tests are necessary for the given product/products? Under which conditions would it be possible to refrain from these biocompatibility tests?

**A:**

Biocompatibility tests are necessary in any case, as IOL are implants. But if these tests were already performed and documented for comparable medical devices, the company could refrain from performing biocompatibility tests. One has to pay attention that the period of application as well as the part of the body where the material is applied, is identical to the tested material.

**Q: Microbiological safety, Material of animal origin**

Describe which aspects are covered in this chapter.

**A:**

Especially with medical devices with material of animal origin:

Viral and bacterial safety, safety according to prions, see literature:

- EN 22442 part 1-3 (Animal tissue and their derivatives used in the manufacture of medical devices, 23a_Final_Case_Task_Ff.docx (http://ec.europa.eu/enterprise/policies/european-standards/harmonised-standards/medical-devices/index_en.htm)

- MEDDEV 2.5-8 (Guidelines on evaluation of medical devices incorporating materials of animal origin with respect to viruses and transmissible agents, http://www.meddev.info/_documents/2_5-8____02-1999.pdf

- Directive 2003/32/EC (applies to material from bovine, ovine, caprine species, deer, elk, cat, and mink only, http://eur-lex.europa.eu/LexUriServ/LexUriServ.do? uri=OJ:L:2003:105:0018:0023:EN:PDF)

The manufacturer has to take the following aspects into consideration and has to provide relevant documents: Justification for the usage of material of animal origin or animal tissue or derivates …

- Basic material
- Species from which the basic material is derived from
- Assessment of the clinical benefit versus remaining risks, especially in comparison to alternative material
- Studies on the elimination or inactivation of BSE/TSE relevant substances
- Applied risk analysis
- Applied measures on the minimization of infection risks
- Control of the basic material, end product and supplier
- Planned audit measures (suppliers of raw material ...)
- EDQM certificate (European Directorate for the Quality of Medicines & Health Care

**Q:**

The viscoelastic that improves the mobility of the implantation system and fosters the safe implantation of the IOL into the eye is a hyaluronic acid product. R&D can purchase this hyaluronic acid from two different suppliers. The first supplier produces hyaluronic acid by microbial fermentation. The second supplier produces hyaluronic acid from raw material that is made from rooster combs. Assess both alternatives as far as safety, classification of the products, long-term reliability of the supplier, extent of the required technical documentation for the registration, the registration process, effects on the registration in other countries (outside of the EU), required manufacturing licenses. Please give a clear recommendation for the management. Keep in mind that viscoelastics are classified in some markets as medicinal product.

**A:**

Assumption 1: Hyaluronic acid is derived from rooster comb > Class III

Rule 17: All devices manufactured utilizing animal tissues or derivatives rendered non-viable are class III except where such devices are intended to come into contact with intact skin only.

Assumption 2: Hyaluronic acid from fermentation without using material of animal origin > class IIb

Rule 8: All implantable devices and long-term surgically invasive devices are in class IIb unless ...

In both cases a technical documentation has to be set up that is comparable as far as the content and the expense is concerned.

The documentation for the class III product has to be assessed by the Notified Body before the CE certification (Design Dossier Review). Some countries assess products of animal origin much more critically than products without material of animal origin. These product may be classified as medicinal products. This would result in significant financial additional expenses, in extensive clinical studies as prerequisite for a registration as well as in a delay in getting the registration of several years.

The source of the basic material of animal origin has to be well controlled on a continuous basis. Outbreaks of epidemics or the toughening of regulations on biological or viral safety is a risk.

As far as the manufacturing license is concerned, there is no relevant difference in the requirements for both procedures.

**Q: Summary**

It is recommended that the basic material derived from fermentation should be used (if both materials are comparable in quality and product characteristics) – especially with the potential risk of a toughening of regulations on the biological or viral safety. Moreover, medical devices with components of animal origin can only be registered globally at significant additional expenses.

Coated medical devices

Is this chapter of the technical documentation relevant for the given product/products? If YES, which aspects should be covered in this chapter?

**A:**

This chapter is not relevant in this study case and will not be part of the technical documentation.

**Q: Clinical data**

The company manufactures IOLs already for centuries and therefore has extensive clinical data. Compared to the CE-certified previous models, the new IOL does not differ in indication, material or design. What's new is the combination of the IOL, viscoelastic and implantation system. Is – from your point of view – a new clinical study necessary? If YES: which aspects have to be covered by the study? Does a clinical assessment have to be conducted or can the company fall back on the updated clinical assessment of the previous IOL? Please explain your recommendation extensively.

**A:**

A clinical assessment has to be performed in any case. Whether or not the company has to conduct a clinical study should be discussed with the Notified Body.

If all potential risks of the pre-loaded configuration can be clinically assessed with existing preclinical studies, references to clinical studies with comparable products (the previous models of the single components), references to market surveillance data of these previous models or via other literature, a new clinical study might not be necessary. It is important to note that Directive 93/42/EEC explicitly requests clinical studies for class III products and implants. Companies can refrain from these studies only in exceptional, clearly warranted cases.

It is recommended to prepare a position paper and to discuss this with the Notified Body.

---

**Q: Qualification of the packaging and shelf life**

Which tests should be carried out and documented in this chapter?

The Marketing department asked for extensive testing while the product is already dispatched in its packaging to various target countries. The integrity of the packaging and the product will be assessed in the target countries.

The Sales Manager who is responsible for Saudi Arabia even asks that the product is stored at a temperature of over 40° Celsius for two weeks and then to assess the integrity of the product and product quality. He argues that the products sometimes stay in customs at Abu Dhabi before being transferred to the selling organization.

When the Sales Manager for Spain and Portugal hears this, he asks the product to be stored for 3 – 4 weeks at 60° Celsius because the products are transported by truck to Spain and Portugal. This transport can last for 3 to 4 weeks, and in summer it could easily be about 60° Celsius.

The head of R&D argues that he has performed these tests already with the similar previous model. He will provide a written justification that these tests can be also used for the new product. One hundred thousand Euro could be saved by this and the product could be placed on the market much earlier.

Assess the different points of view and give a recommendation for the management how to proceed.

**A:**

Full account shall be taken of the examination and assessment of the stability and integrity of the product packaging. Relevant requirements should be applied. In case that the product is exposed to high temperatures, this has to be tested. The sterility and the product performance have to be ensured even after transport and storage under real-life conditions until the expiry date. A repeated testing may not be necessary under certain circumstances, i. e. if tests have been performed with a comparable product with a comparable packaging under relevant circumstances. The company could refer to these tests and reports.

**Q: Product labelling**

List the mandatory information that has to be printed on the product labelling.

**A:**

According to Directive 93/42/EEC Annex I, 13, the following information has to be supplied by the manufacturer:

13.1. Each device must be accompanied by the information needed to use it safely and properly, taking account of the training and knowledge of the potential users, and to identify the manufacturer. This information comprises the details on the label and the data in the instructions for use. As far as practical and appropriate, the information needed to use the device safely must be set out on the device itself and/or on the packaging for each unit or, where appropriate, on the sales packaging. If individual packaging of each unit is not practical, the information must be set out in the leaflet supplied with one or more devices. Instructions for use must be included in the packaging for every device. By way of exception, no such instructions for use are needed for devices of class I or IIa if they can be used safely without any such instructions.

13.2. Where appropriate, this information should take the form of symbols. Any symbol or identification color used must conform to the harmonized standards. In areas for which no standards exist, the symbols and colors must be described in the documentation supplied with the device.

13.3. The label must bear the following particulars:

a) the name or trade name and address of the manufacturer. For devices imported into the Community, in view of their distribution in the Community, the label, or the outer packaging, or instructions for use, shall contain in addition the name and address of the authorised representative where the manufacturer does not have a registered place of business in the Community;

b) the details strictly necessary to identify the device and the contents of the packaging especially for the users;

c) where appropriate, the word "STERILE";

d) where appropriate, the batch code, preceded by the word "LOT", or the serial number;

e) where appropriate, an indication of the date by which the device should be used, in safety, expressed as the year and month;

f) where appropriate, an indication that the device is for single use. A manufacturer's indication of single use must be consistent across the Community;

g) if the device is custom-made, the words "custom-made device" (-> not applicable in this study case!)

h) if the device is intended for clinical investigation, the words "exclusively for clinical investigations" (-> not applicable in this study case!)

i) any special storage and/or handling conditions;

j) any special operating instructions;

k) any warnings and/or precautions to take;

l) year of manufacture for active devices other than those covered by (e). This indication may be included in the batch or serial number (-> not applicable in this study case!)

m) where applicable, method of sterilization;

n) in the case of a device in the meaning of Article 1(4a), an indication that the device contains a human blood derivative (-> not applicable in this study case!).

13.4. If the intended purpose of the device is not obvious to the user, the manufacturer must clearly state it on the label and in the instructions for use.

13.5. Wherever reasonable and practicable, the devices and detachable components must be identified, where appropriate in terms of batches, to allow all appropriate action to detect any potential risk posed by the devices and detachable components (-> not applicable in this study case!)

13.6 Where appropriate, the instructions for use must contain the following particulars:

a) the details referred to in Section 13.3, with the exception of (d) and (e);

b) the performances referred to in Section 3 and any undesirable side effects;

c) if the device must be installed with or connected to other medical devices or equipment in order to operate as required for its intended purpose, sufficient details of its characteristics to identify the correct devices or equipment to use in order to obtain a safe combination;

d) all the information needed to verify whether the device is properly installed and can operate correctly and safely, plus details of the nature and frequency of the maintenance and calibration needed to ensure that the devices can operate properly and safely at all times;

e) where appropriate, information to avoid certain risks in connection with implantation of the device;

f) information regarding the risks of reciprocal interference posed by the presence of the device during specific investigations or treatment;

g) the necessary instructions in the event of damage to the sterile packaging and, where appropriate, details of appropriate methods of resterilisation;

h) if the device is reusable ... will still comply with the requirements in Section I > not applicable in the study case (single use device). If the device bears an indication that the device is for single use, information on known characteristics and technical factors known to the manufacturer that could pose a risk if the device were to be re-used. If in accordance with Section 13.1. no instructions for use are needed, the information must be made available to the user upon request (-> not applicable in this study case!)

i) details on any further treatment or handling needed before the device can be used, i.e. sterilization, final assembly ), (-> not applicable in this study case!)

j) in the case of devices emitting radiation (-> not applicable in this case!)

The instructions for use must also include all details allowing the medical staff to brief the patient on any contra-indications and any precautions to be taken. These details should cover in particular:

k) precautions to be taken in the event of changes in the performance of the device;

l) precautions to be taken as regards exposure, in reasonably foreseeable environmental conditions, to magnetic fields, external electrical influences, electrostatic discharge, pressure or variations in pressure, acceleration, thermal ignition sources, etc;

m) adequate information regarding the medicinal product or products which the device in question is designed to administer, including any limitations in the choice of substances to be delivered;

n) precautions to be taken against any special, unusual risks related to the disposal of the device;

o) medicinal substances, or human blood derivatives incorporated into the device as an integral part in accordance with Section 7.4 > not applicable in this study case!

p) degree of accuracy claimed for the devices with a measuring function > not applicable in this study case!

q) date of issue or the latest revision of the instructions for use.

**Q:** The Product Manager does not want a big label with all mandatory information because this would disrupt the looks of the packaging. He wants this information to be printed on several small labels. What do you think about this?

**A:**
This can be agreed to as there is no requirement that all information has to be on one label.

**Q:** The Sales Manager for Spain and Portugal wants the name of the product (that defines the kind of the product) to be translated in Spanish and Portuguese. The head of production opposes this and mentions that all mandatory information is covered by EN 980 symbols and therefore translations are not necessary. Do you agree?

**A:**

The use of symbols is explicitly welcomed in Directive 93/42/EEC. Information that are covered by symbols of the harmonized norm EN 980 do not have to be translated. Their meaning does not have to be explained in an official European language. If it is not obvious, it needs to be explained what kind of a product it is. This could be done via a self-developed symbol. Its meaning must be explained in the instructions for use in all languages of the target countries. Alternatively, the precise declaration of the product can also be printed in words on the label and consequently translated in all languages of the target countries.

**Q:** Promotional material

The main European congress in this field will take place in autumn. It is absolutely essential for the company to present and to promote the product at this congress. At the moment it looks like the registration approval will be received several weeks after the congress. The Sales Manager as well as the Head of Marketing want to actively promote the product at this congress with posters and promotional videos which show the product being applied. Moreover, Marketing wants to give product samples to potential customers. The Head of Clinical Research has agreed to have a live operation (where the product is used) from the operation theatre broadcasted to the congress.

Assess these proposals taking into account the registration approval is not likely to be received at the time of the congress. Please give a recommendation. Is it allowed to present not registered products at a congress? If YES: what are the prerequisites? Is it allowed to give samples to potential customers? Is it possible to give out non-sterile products that are labelled accordingly? What about the broadcast from the operation theatre? Could you think of a scenario where this would be possible?

**A:**

The regulations regarding promotion for medical devices differ from country to country. Therefore it is key in which country the congress will take place because the applicable law of this country has to be complied with. It is allowed, in general, to display (at a congress) medical devices that are not yet CE certified. Of course, on all information material it has to be stated that the product is not CE certified. Sample products have to be unsterile and have to be clearly labelled as "not for use on human beings". Comparative advertising for non-CE certified medical devices is not allowed.

The surgery with the non-CE certified medical device can only take place within the scope of a clinical study. The positive vote of the Ethics Committee and the consent from the patient is necessary. All documents needed for a clinical study and all necessary permissions have to be available.

---

**Q:** Sterilization

Which documents on sterilization are usually part of the technical documentation?

**A:**

Information on sterilization can be found in EN 550 series and ISO 11135 series. Description of the qualification of the sterilization process as well as the summary of the validation (the chosen method must achieve a SAL (Sterility Assurance Level) of $10^{-6}$). Process validation (report) including a physical and microbial performance records. Proof of the certification of the sterilization facilities by a Notified Body (ISO 9001/2, ISO 13485/13488, EN 550 series, ISO 11135 series).

---

**Q:** Final evaluation

The risk of the product and of its application has been reduced by various iterative processes. Nevertheless, the final evaluation shows a residual risk. Is it possible to have the product registered nevertheless? What has to be the conclusion of the final evaluation?

**A:**

A product with a residual risk can also be CE certified. It is important that the residual risk can be reduced as much as possible and that the manufacturer comes to the conclusion that the medical benefit outweighs the residual risk.

**Q:** Declaration of conformity by the manufacturer

Draw up a draft of the declaration of conformity for the given product/products

**A:**

please see next page ->

## EC Declaration of Conformity

We

*-owner of the CE certificate incl. the address of the manufacturer-*

declare on our own responsibility that the product *"xyz"* and associated accessories, meet all the provisions of the Directive 93/42/EEC which apply to them.

*-Notified Body and address-*

Conformity assessment procedure
Annex _, Section _ of the Directive 93/42/EEC.

EU product class _
Class _, Rule _ of Annex IX of MDD

EC certificate number
*-EC Certificate Number-*

This document is valid from the day of signature until _ (date i. e. expiry date of the relevant CE certificate)

Date and signature

**Annex 1:**

**Essential Requirements**

PRELOADED DELIVERY SYSTEM CONTAINING 1-PIECE SOFT ACRYLIC LENSES

Vision 1-Piece lens as Model VCB00

## MDD 93/42/EEC - Annex 1

### General Requirements

| No. | Essential Requirement | Proof of compliance | Reference to document |
|-----|-----------------------|---------------------|------------------------|
| 1. | The device must be designed and manufactured in such a way that, when used under the conditions and for the purposes intended, they will not compromise the clinical condition or the safety of patients, or the safety and health of users or, where applicable, other persons, provided that any risks which may be associated with their use constitute acceptable risks when weighed against the benefits to the patient and are compatible with a high level of protection of health and safety. | | |
| 2. | The solutions adopted by the manufacturer for the design and construction of the devices must conform to safety principles, taking account of the generally acknowledged state of the art. In selecting the most appropriate solutions, the manufacturer must apply the following principles in the following order: - Eliminate or reduce risks as far as possible (inherently safe design and construction) - Where appropriate take adequate protection measures including alarms if necessary, in relation to risks that can not be eliminated. - inform users of the residual risks due to shortcomings of the protection methods adopted. | | |
| 3. | The devices must achieve the performance intended by the manufacturer and be designed, manufactured and packaged in such a way that they are suitable for one or more of the functions referred | | |

to in Article 1 (2) (a) as specified by the manufacturer.

**Essential Requirements: Medical Device Directive 93/42/EEC - Annex I**

| | | | |
|---|---|---|---|
| 4. | The characteristics and performance referred to in sections 1, 2 and 3 must not be adversely affected to such a degree that the clinical condition and safety of the patients and, where applicable, of other persons are compromised during the lifetime of the device as indicated by the manufacturer, when the device is subjected to the stresses which can occur during normal conditions of use. | | |
| 5. | The devices must be designed, manufactured and packed in such a way that their characteristics and performances during their intended use will not be adversely affected during transport and storage taking into account the instructions and information provided by the manufacturer. | | |
| 6. | Any undesirable side effects must constitute an acceptable risk when weighed against the performances intended. | | |

## II. REQUIREMENTS REGARDING DESIGN AND CONSTRUCTION

| No. | Essential Requirement | Proof of compliance | Reference to document |
|---|---|---|---|
| 7. | Chemical, Physical and Biological Properties | | |
| 7.1. | The devices must be designed, manufactured and packed in such a way as to guarantee the characteristics and performances referred to in Section I on the "General Requirements". Particular attention must be paid to: | | |

| | | | |
|---|---|---|---|
| | - the choice of materials used, particularly as regards toxicity and, where appropriate, flammability<br>- the compatibility between the materials used and biological tissues, cells and body fluids, taking account of the intended purpose of the device. | | |
| 7.2. | The devices must be designed, manufactured and packed in such a way as to minimize the risk posed by contaminants and residues to the persons involved in the transport, storage and use of the devices and to the patients, taking account of the intended purpose of the product. Particular attention must be paid to the tissues exposed and the duration and frequency of the exposure. | | |

## Essential Requirements, Medical Device Directive 93/42/EEC - Annex I

| | | | |
|---|---|---|---|
| 7.3. | The devices must be designed and manufactured in such a way that they can be used safely with the materials, substances and gases with which they enter into contact during normal use or during routine procedures; if the devices are intended to administer medicinal products, they must be designed and manufactured in such a way as to be compatible with the medicinal products concerned according to the provisions and restrictions governing those products and that their performance is maintained in accordance with the intended use. | | |
| 7.4. | Where a device incorporates, as an integral part, a substance which, if used separately, may be considered to be a medicinal product as defined in Article I of Directive 65/65/EEC and which is liable to act upon the body with action ancillary to that of the device, the safety, quality and usefulness of the substance must be verified, taking account of the intended purpose of the device, by analogy with the appropriate methods specified in Directive 75/318/EEC. | | |

| | | | |
|---|---|---|---|
| 7.5. | The devices must be designed and manufactured in such a way as to reduce to a minimum the risks posed by substances leaking from the device. | | |
| 7.6. | The devices must be designed and manufactured in such a way as to reduce as much as possible, risks posed by the unintentional ingress of substances into the device taking into account the device and the nature of the environment in which it is intended to be used. | | |

**Essential Requirements, Medical Device Directive 93/42/EEC - Annex I**

| No. | Essential Requirement | Proof of compliance | Reference to document |
|---|---|---|---|
| 8. | **Infection and Microbial Contamination** | | |
| 8.1. | The devices and their manufacturing processes must be designed in such a way as to eliminate or reduce as far as possible the risk of infection to the patient, user and third parties, the design must allow easy handling and, where necessary, minimize contamination of the device by the patient or vice versa during use. | | |

**Essential Requirements, Medical Device Directive 93/42/EEC - Annex I**

| No. | Essential Requirement | Proof of compliance | Reference to document |
|---|---|---|---|
| 8.2. | Tissues of animal origin must originate from animals that have been subjected to veterinary controls and surveillance adapted to the intended use of the tissues. Notified Bodies shall retain information on the geographical origin of the animals. | | |

| | | | |
|---|---|---|---|
| | Processing, preservation, testing and handling of tissues, cells and substances of animal origin must be carried out so as to provide optimal security. In particular, safety with regard to viruses and other transferable agents must be addressed by implementation of validated methods of elimination or viral inactivation in the course of the manufacturing process. | | |
| 8.3. | Devices delivered in a sterile state must be designed, manufactured and packed in a non-reusable pack and/or according to appropriate procedures to ensure they are sterile when placed on the market and remain sterile, under the storage and transport conditions laid down, until the protective packaging is damaged or opened. | | |
| 8.4. | Devices delivered in a sterile state must have been manufactured and sterilized by an appropriate, validated method. | | |
| 8.5. | Devices intended to be sterilized, must be manufactured in appropriately controlled (e.g. environmental) conditions. | | |
| 8.6. | Packaging systems for non-sterile devices must keep the product without deterioration in the level of cleanliness stipulated and, if the devices are to be sterilized prior to use, minimize the risk of microbial contamination. The packaging system must be suitable taking account of the method of sterilization indicated by the manufacturer. | | |
| 8.7. | The packaging and/or label of the device must distinguish between identical or similar products sold in both sterile and non-sterile condition. | | |

**Essential Requirements, Medical Device Directive 93/42/EEC - Annex I**

| No. | Essential Requirement | Proof of compliance | Reference to document |
|---|---|---|---|
| 9. | **Construction and Environmental Properties** | | |
| 9.1. | If the device is intended for use in combination with other devices or equipment, the whole combination, including the connection system must be safe and must not impair the specified performance of the devices. Any restrictions on the use must be indicated on the label or in the instructions for use. | | |
| 9.2. | Devices must be designed and manufactured in such a way as to remove or minimize as far as possible:<br>- the risk of injury, in connection with their physical features, including the volume/pressure ratio, dimensional, and where appropriate the ergonomic features,<br>- risks connected with reasonable foreseeable environmental conditions, such as magnetic fields, external electrical influences, electrostatic discharge, pressure, temperature or variations in pressure and acceleration,<br>- the risks of reciprocal interference with other devices normally used in the investigations or for the treatment given,<br>- risks arising where maintenance or calibration are not possible (as with implants) from ageing of the materials used or loss of accuracy of any measuring or control mechanism. | | |

| 9.3. | Devices must be designed and manufactured in such a way as to minimize the risks of fire or explosion during normal use and in single fault condition. Particular attention must be paid to devices whose intended use includes exposure to flammable substances which could cause combustion. | | |
|---|---|---|---|

**Essential Requirements, Medical Device Directive 93/42/EEC - Annex I**

| No. | Essential Requirement | Proof of compliance | Reference to document |
|---|---|---|---|
| 10. | Devices with measuring function. | | |

**Essential Requirements, Medical Device Directive 93/42/EEC - Annex I**

| No. | Essential Requirement | Proof of compliance | Reference to document |
|---|---|---|---|
| 11. | Protection against radiation | | |

**Essential Requirements, Medical Device Directive 93/42/EEC - Annex I**

| No. | Essential Requirement | Proof of compliance | Reference to document |
|---|---|---|---|
| 12. | Requirements for medical devices connected to or equipped with an energy source | | |

## Essential Requirements, Medical Device Directive 93/42/EEC - Annex I

| No. | Essential Requirement | Proof of compliance | Reference to document |
|-----|----------------------|---------------------|----------------------|
| 13. | **Information supplied by the manufacturer** | | |
| 13.1. | Each device must be accompanied by the information needed to use it safely and to identify the manufacturer, taking account of the training and knowledge of the potential user. This information comprises the details on the label and the data in the instructions for use. As far as practicable and appropriate, the information needed to use the device safely must be set out on the device itself and/or on the packaging of each unit. If not practicable, the information must be set out in the leaflet supplied with one or more devices. Instructions for use must be included in the packaging for every device. By way of exception, no such instruction leaflet is needed for devices in Class I or Class IIa if they can be used completely safely without any such instructions. | | |
| 13.2. | Where appropriate, this information should take the form of symbols. Any symbol or identification colour used must conform to the harmonized standards. In areas for which no standards exist, the symbols and colours must be described in the documentation supplied with the device. | | |

**Essential Requirements, Medical Device Directive 93/42/EEC - Annex I**

| No. | Essential Requirement | Proof of compliance | Reference to document |
|-----|----------------------|---------------------|----------------------|
| 13.3. | The label must bear the following particulars:<br>a. the name or trade name and address of the manufacturer. For devices imported into the Community, the label, or the outer packaging, or the instructions for use, shall contain in addition the name and address of either the person responsible referred to in Article 14.2 or of the authorized representative of the manufacturer established within the Community or of the importer established within the Community, as appropriate.<br>b. the details strictly necessary for the user to identify the device and the contents of the packaging.<br>c. where appropriate, the word 'STERILE'.<br><br>d. where appropriate, the batch code, preceded by the word 'LOT' or the serial number.<br>e. where appropriate, an indication of the date by which the device should be used, in safety, expressed as the year and month.<br>f. where appropriate, an indication that the device is for single use.<br>g. if the device is custom-made, the words "custom made device".<br>h. if the device is intended for clinical investigations, the words "exclusively for clinical investigations".<br>i. any special storage and/or handling conditions.<br>j. any special operating instructions. | | |

    k. any warnings and/or precautions to take.

    l. year of manufacture of active devices, other than those covered by e). This indication may be included in the batch or serial number.

    m. where applicable, method of sterilization.

13.4. If the intended purpose of the device is not obvious to the user, the manufacturer must clearly state it on the label and in the instructions for use.

13.5. Wherever reasonable and practicable, the devices and detachable components must be identified, where appropriate in terms of batches, to allow all appropriate action to detect any potential risk posed by the devices and detachable components.

13.6. Where appropriate, the instructions for use must contain the following particulars:

    n. the details referred to in 13.3, with the exception of d) and e).

    o. The performances referred to in Section 3 and any undesirable side effects.

    p. If the device must be installed with or connected to other medical devices or equipment in order to operate as required for its intended purpose, sufficient details of its characteristics to identify the correct devices or equipment to use in order to obtain a safe combination.

    q. All the information needed to verify whether the device is properly installed and can operate correctly and safely, plus details of the nature and frequency of the maintenance and calibration needed to ensure that the devices operate properly and safely at all times.

    r. Where appropriate, information to avoid certain risks in connection with implantation of the device.

    s. Information regarding the risks of reciprocal interference posed by the presence of the device during specific investigations or treatment.

t. The necessary instructions in the event of damage to the sterile packaging and where appropriate, details or appropriate methods of re-sterilization.

u. If the device is reusable, information on the appropriate processes to allow reuse, including cleaning, disinfection, packaging, and where appropriate, the method of sterilization of the device to be re-sterilized and any restriction on the number of uses.

v. Where the devices are supplied with the intention that they be sterilized before use, the instructions for cleaning and sterilization must be such that, if correctly followed,, the device will still comply with the requirements in Section 1.

w. Details of any further treatment or handling needed before the device can be used (for example sterilization, final assembly etc.)

x. In the case of devices emitting radiation for medical purposes, details of the nature, type, intensity and distribution of this radiation. The instructions for use must include details allowing the medical staff to brief the patient on any contra-indications and any precautions to be taken. These details should cover in particular:

y. Precautions to be taken in the event of changes in the performance of the device.

z. Precautions to be taken as regards exposure, in reasonable foreseeable environmental conditions, to magnetic fields, external electrical influences, electrostatic discharge, pressure or variations in pressure, acceleration, thermal ignition sources, etc.

aa. Adequate information regarding the medicinal products which the device in question is designed to administer, including any

limitations in the choice of substances to be delivered.
a. precautions to be taken against any special, unusual risks related to the disposal of the device
b. medicinal substances incorporated into the device as an integral part in accordance with Section 7.4
c. degree of accuracy claimed for devices with a measuring function.

14. Where conformity with the essential requirements must be based on clinical data, as in Section 1 (6), such data must be established in accordance with Annex X.

**Annex 2:**

**LIST OF STANDARDS -** PRELOADED DELIVERY SYSTEM CONTAINING 1-PIECE SOFT ACRYLIC LENSES

Vision 1-Piece lens as Model VCB00

| STANDARD REVISION LIST | | |
|---|---|---|
| **NUMBER** | **REVISION DATE** | **TITLE** |
| Declaration of Helsinki | 2013 | World Medical Association Declaration of Helsinki – Ethical Principles for Medical Research Involving Human Subjects |
| EN 980 | 2008 | Symbols for Use in the Labeling of Medical Devices |
| EN 15223-1 | 2012 | Medical devices -- Symbols to be used with medical device labels, labelling and information to be supplied -- Part 1: General requirements |
| EN 1041 | 2008 | Information Supplied by the Manufacturer with Medical Devices |
| EN 62366 | 2008 | Medical Devices – Application of Usability Engineering to Medical Devices |
| EN ISO 10993-1 | 2009 | Biological Evaluation of Medical Devices – Part 1: Evaluation and Testing Within a Risk Management Process |
| EN ISO 10993-3 | 2003 | Biological Evaluation of Medical Devices – Part 3: Test for Genotoxicity, Carcinogenicity and Reproductive Toxicity |
| EN ISO 10993-5 | 2009 | Biological Evaluation of Medical Devices – Part 5: Test for Cytotoxicity, In Vitro Methods |
| EN ISO 10993-6 | 2009 | Biological Evaluation of Medical Devices – Part 6: Test for Local Effects After Implantation |
| EN ISO 10993-7 | 2008 | Biological Evaluation of Medical Devices – Part 7: Ethylene Oxide Sterilization Residuals |
| EN ISO 10993-10 | 2010 | Biological Evaluation of Medical Devices – Part 10: Test for Irritation and Sensitization |
| EN ISO 10993-11 | 2009 | Biological Evaluation of Medical Devices – Part 11: Test for Systemic Toxicity |
| EN ISO 11135-1 | 2007 | Sterilization of Health Care Products – Ethylene Oxide – Part 1 : Development, Validation and Routine Control of a Sterilization Process for Medical Devices |
| ISO/TS 11135-2 | 2008 | Sterilization of Health Care Products – Ethylene Oxide – Part 2: Guidance on the Application of ISO 11135-1 |
| ISO 11138-1 | 2006 | Sterilization of Health Care Products – Biological Indicators – Part 1: General Requirements |
| EN ISO 11138-2 | 2006 | Sterilization of Health Care Products – Biological Indicators – Part 2: Biological Indicators for Ethylene Oxide Sterilization Processes |

## STANDARD REVISION LIST

| NUMBER | REVISION DATE | TITLE |
|---|---|---|
| EN ISO 11607-1 | 2009 | International Standard – Packaging for Terminally Sterilized Medical Devices – Part 1: Requirements for Materials, Sterile Barrier Systems, and Packaging |
| EN ISO 11607-2 | 2006 | International Standard – Packaging for Terminally Sterilized Medical Devices – Part 2: Validation Requirements for Forming, Sealing and Assembly Process |
| EN ISO 11737-1 | 2006 | Sterilization of Medical Devices – Microbiological Methods – Part 1: Determination of a Population of Microorganisms on Products |
| EN ISO 11737-2 | 2009 | Sterilization of Medical Devices – Microbiological Methods – Part 2: Tests of Sterility Performed in the Validation of a Sterilization Process |
| ISO 11979-1 | 2006 | Ophthalmic Implants – Intraocular Lenses – Part 1: Vocabulary |
| ISO 11979-2 | 1999 | Ophthalmic Implants – Intraocular Lenses – Part 2: Optical Properties and Test Methods |
| ISO 11979-3 | 2006 | Ophthalmic Implants – Intraocular Lenses – Part 3: Mechanical Properties and Test Methods |
| ISO 11979-4 | 2008 | Ophthalmic Implants – Intraocular Lenses – Part 4: Labeling and Information |
| ISO 11979-5 | 2006 | Ophthalmic Implants – Intraocular Lenses – Part 5: Biocompatibility |
| ISO 11979-6 | 2007 | Ophthalmic Implants – Intraocular Lenses – Part 6: Shelf Life and Transport Stability |
| ISO 11979-7 | 2006 | Ophthalmic Implants – Intraocular Lenses – Part 7: Clinical Investigations |
| EN ISO 11979-8 | 2009 | Ophthalmic Implants – Intraocular Lenses – Part 8: Fundamental Requirements |
| EN ISO 13485 | 2003 | Quality Systems – Medical Devices – System Requirements for Regulatory Purposes |
| ISO 14155 | 2011 | Clinical Investigation of Medical Devices for Human Subjects – Good Clinical Practices |
| ISO 14644-1 | 1999 | Cleanrooms and Associated Controlled Environments – Part 1: Classification of Air Cleanliness |
| ISO 14644-2 | 2000 | Cleanrooms and Associated Controlled Environments – Part 2: Specifications for Testing and Monitoring to Prove Continuous Compliance with ISO 14644-1 |
| EN ISO 14971 | 2012 | Medical Devices – Application of Risk Management to Medical Devices |
| ISO 15223-1 | 2007 | Medical Devices – Symbols to be Used with Medical Device Labels, Labeling and Information to be Supplied – Part 1: General Requirements |
| ISO 15223-1 Amendement 1 | 2008 | Medical Devices – Symbols to be Used with Medical Device Labels, Labeling and Information to be Supplied – Part 1: General Requirements Amendment 1 |

| STANDARD REVISION LIST | | |
|---|---|---|
| **NUMBER** | **REVISION DATE** | **TITLE** |
| ISTA-2A | 2011 | Packaged Products 150 lb (68 kg) or Less |
| EN ISO 22442-1 | 2007 | Medical devices utlizing animal tissues and their derivatives. Analysis and management of risk |
| EN ISO 22442-2 | 2007 | Medical devices utlizing animal tissues and their derivatives. Controls on sourcing, collection and handling |
| EN ISO 22442-3 | 2007 | Medical devices utlizing animal tissues and their derivatives. Validation of the elimination and/or inactivation of viruses and transmissible agents |

## Chapter 24: Glossary

*Dr. Stefan Menzl, Dr. Sibylle Scholtz, Dr. Carsten Rupprath,*

*Myriam Becker*

| | |
|---|---|
| AIMD | Active Implantable Medical Device |
| ANSM | L'Agence nationale de sécurité du médicament et des produits de santé; French Competent Authority (since 1st May 2012, formerly known as "AFSSAPS"). |
| ATMP | Advanced Therapy Medicinal Product; new therapies; umbrella term for three classes of medicinal product groups:<br><br>- somatic cell therapy<br>- gene therapy<br>- tissue engineering<br><br>These medicinal products often contain or exist of viable cells or tissue and thus are characterized by high complexity. Often the used cells are emanated from a patient, processed in a laboratory, (e.g. bred or genetically modified), and subsequently administered the same patient. ATMPs are therefore often an example for "personalized medicine". |
| Audit | An audit (latin audire: listen; audit, he/she/it listens; also translated as hearing/consultation) is generally described as analytical methods which serve to assess processes related to meeting requirements and directives. Often this takes place in the scope of Quality Management. The audits are being executed by a specially trained auditor. |

| | |
|---|---|
| Authorized European Representative (EC REP) | Contact person of a manufacturer of medical devices in Europe; this representative is required when the manufacturer does not have a place of business in Europe, however wants to distribute products there. The EC REP acts as contact person for the national authorities in Europe. |
| BfArM | Bundesinstitut für Arzneimittel und Medizinprodukte (Federal Institute for Drugs and Medical Devices; it is an independent higher federal authority within the portfolio of the Federal Ministry of Health. |
| CE (Kennzeichnung) | French: Conformité Européenne (European Conformity); The CE marking is a symbol of free marketability in the European Economic Area. CE marking signifies that the product conforms with all EC directives that apply to it, fulfills the Essential Requirements and has passed successfully the mandatory conformity assessments. |
| CEN | European Committee for Standardization |
| CENELEC | European Committee for Electrotechnical Standardization |
| Classification of medical devices | Medical devices, with the exemption of active implantable medical devices, are classified in certain classes. The classification is done according to the rules of annex IX of directive 93/42/EEC. In vitro diagnostic medical devices are classed according to a category of annex II of directive 98/79/EC or are devices for self-testing or other devices. |

| Compliance | Adherence to statutory requirements and regulatory standards, but also fulfilment of a company to set ethical standards and requirements. |
| --- | --- |
| Conformity | Compliance with statutory requirements (related to safety and performance) of a medical device directive; the manufacturer must confirm compliance with the conformity assessment before allowed to place the product on the market. |
| Conformity assessment procedure | Method selected by the manufacturer in order to demonstrate conformity of compliance with regulatory requirements (according to one of the medical device directives). Depending on the risk class of a medical device various procedures are possible which can be obtained from the annexes of the respective medical device directive. |
| DAkkS | Deutsche Akkreditierungsstelle GmbH; national accreditation body for the Federal Republic of Germany, with headquarters in Berlin. It assesses, attests and monitors the technical competence of laboratories, certification and inspection bodies as an independent body. In this context, DAkkS makes an important contribution to the quality assurance of products and services, consumer trust and the competitiveness of the German economy. |
| Declaration of Conformity (DoC) | Result of a successful conformity assessment; it is the written statement and the single declaration drawn up by the manufacturer to demonstrate the fulfillment of the EU directives (AIMD, MDD, IVDD) relating to a product bearing the CE |

| | marking he has manufactured. |
|---|---|
| Designation | Figuratively, the registered place of business of the manufacturer. |
| DG | Directorate General; department of the European Commission, classified according to the policy it deals with, e.g. Competition, Energy, Health and Consumers (SANCO) Justice. |
| DG SANCO | Directorate General for Health and Consumer Protection |
| Directive | E. g., the three core directives for medical devices:<br><br>- Directive on Active Implantable Medical Devices (AIMD), 90/385/EEC<br>- Medical Device Directive (MDD), 93/42/EEC<br>- Directive on in vitro diagnostic medical devices (IVDD), 98/79/EC |
| EC | European Community |
| EDMA | European Diagnostic Manufacturers Association; is an international, non-profit organization representing the interests of the medical in vitro diagnostics industry in Europe |
| EEA | European Economic Area; enlargement of the European economic zone with those countries (Iceland, Liechtenstein, Norway) that do not pertain to the European Union or EFTA. It is based on mutual recognition between the EFTA states and the European Union. |
| EEC | European Economic Community |

| EFTA states | European Free Trade Association; consists of four member states (Iceland, Liechtenstein, Norway and Switzerland) which are closely related to the European Union. Switzerland has a bilateral agreement with the European Union. |
|---|---|
| Essential Requirements | Minimum requirements which a medical device manufacturer has to fulfill according to the three medical device directives. |
| ETSI | European Telecommunications Standards Institute |
| Global Approach | Serves the removal of barriers to trade and ensures the free movement of goods within the European Union; it is based on harmonized standards; with conformity of a product to these standards it can be considered that the Essential Requirements have been fulfilled. |
| Harmonised Standards | A harmonized standard is a European standard elaborated on the basis of a request from the European Commission to a recognized European Standards Organization (CEN, CENELEC or ETSI) to develop a European standard that provides solutions for compliance with a legal provision. Such a request provides guidelines which requested standards must respect to meet the essential requirements or other provisions of relevant European Union harmonization legislation. |
| Intended use | A product's *intended use* is what it is supposed to be used for according to the manufacturer's specifications, instructions, and other information, often found in the manufacturer's package insert |

| | |
|---|---|
| | or directions for use. |
| ISO | International Organization for Standardization; develops and published international standards in all areas except electronics and telecommunications. |
| IVD | In vitro diagnostic medical device; definition according to the IVD-Directive: <br><br>"In vitro diagnostic medical device means any medical device which is a reagent, reagent product, calibrator, control material, kit, instrument, apparatus, equipment, or system, whether used alone or in combination, intended by the manufacturer to be used in vitro for the examination of specimens, including blood and tissue donations, derived from the human body, solely or principally for the purpose of providing information: <br><br> - concerning a physiological or pathological state, or <br> - concerning a congenital abnormality, or <br> - to determine the safety and compatibility with potential recipients, or <br> - to monitor therapeutic measures." |
| IVDD | In vitro diagnostic medical device directive; 98/79/EG of the European Parliament and of the Council of 27 October 1998; covers the performance, placing on the market and the CE marking of in vitro diagnostic medical devices in th European Economic Community. |
| MD | Medical Device |

| MDD | Medical Device Directive, 93/42/EEC of 14 June 1993 |
|---|---|
| MDEG | Medical Device Expert Group |
| MEDDEV | Acronym for Medical Devices; collection of consensus documents or guidelines of various working groups of the European Commission for the application of the EC medical device directives. |
| MepV | Acronym for the Swiss "Medizinprodukte-verordnung" (Medical Devices Ordinance) of 17 October 2001 |
| MPG | Acronym for the German and Austrian "Medizin-produktegesetz" (Medical Devices Act). It is the implementation into national legislation of the directives 90/385/EEC, 93/42/EEC and 98/79/EC. |
| MPSV | Acronym for Medizinprodukte-Sicherheitsplan-verordnung (Order on the Medical Devices Vigilance System); order on the recording, assessment and counteractive measure of risks. |
| MPV | Acronym for Medizinprodukte-Verordnung (Medical Device Ordinance); order on conformity assessment. |
| NB-MED | Notified Bodies Medical Devices |
| NBOG | Notified Body Operations Group<br><br>In July 2000, Member States and the European Commission agreed to set up the NBOG. This was in response to widespread concern that the performance of Notified Bodies in the medical device sector, and that of the Designating |

| | Authorities responsible for them, was variable and inconsistent. NBOG membership consists of the European Commission, and nominees from the Member States Designating/Competent Authorities. Additionally, membership of the Group is open to EFTA/EEA Competent Authorities as well as Candidate and Accession countries. NBOG works primarily by the production of written guidance and advice. |
|---|---|
| New Approach/Global Approach | Both concepts were introduced for the removal of barriers to trade whereby the New Approach describes the product regulation and the Global Approach the conformity assessment. It introduces, among other things, a clear separation of responsibilities between the EC legislator and the European standards bodies CEN, CENELEC and ETSI in the legal framework allowing for the free movement of goods. |
| Notified Body | A Notified Body is accredited by a Member State; it carries out conformity assessments laid down in the relevant harmonized European standards or European Technical Assessment on behalf of the medical device manufacturer. |
| OJEC | Official Journal of the European Community |
| Opinion | In connection with "Regulatory opinion": Position paper /comment from a point of view of a "Regulatory Affairs" staff member. |
| PEI | Paul–Ehrlich–Institut; The Paul-Ehrlich-Institut is an institution of the Federal Republic of Germany. It reports to the Bundesministerium für Gesundheit (Federal Ministry of Health). Its |

| | research and control activities promote the quality, efficacy and safety of biological medicinal products |
|---|---|
| Placing on the market | Is related to the responsibility of that person who first places a product on the market. The responsible for placing on the market must be identifiable in case of liability claims or risks connected with a product which were not declared correctly or which did not fulfill the regulatory provisions. |
| PMS | Post-Market Surveillance; monitoring of medical devices placed on the market; based on the evaluation of adverse events and other actions. |
| Presumption of conformity | Presumption of adherence of conformity to one of the medical device directives in "positive" sense. It is presumed that the medical device manufacturer has fulfilled the essential requirements of the regulation (according to AIMD, MDD, IVDD), insofar the manufacturer has designed and manufactured his products according to harmonized standards. |
| Recast/Recast of the medical device directives | Revision of the medical device directives with the aim to guarantee the safety and health protection – also in respect of new products, to allow an innovation-favourable legal framework for a fast market access and to optimize the framework conditions for a working domestic market. |
| Risk classes | Define the classification of a medical device, based on the classification and duration or location of the medical device application (according to AIMD/ MDD) or on the risk a product |

|  | could pose to the user or environment (according to IVDD). The risk class of a medical device determines the conformity assessment procedure for the medical device manufacturer and can be retrieved from the respective directive (AIMDD, MDD, IVDD). |
| --- | --- |
| STED | GHTF SG1 N11: 2008 Summary Technical Documentation for Demonstrating Conformity to the Essential Principles of Safety and Performance of Medical Devices. |
| Team-NB | European Association for Medical Devices of Notified Bodies |
| Treaty of Lisbon | The Treaty of Lisbon was signed on 13 December 2007 and entered into force on 1 December 2009. The Treaty of Lisbon reinforces democracy in the EU and its capacity to promote the interests of its citizens on a day-to-day basis. With the Treaty of Lisbon the EU defined consistent and compulsory requirements in the form of EEC/EC/EU directives so that products meet essential requirements (mainly related to usage and safety). It reformed the Treaty on European Union and the Treaty establishing the European Community to a new "Treaty on the Functioning of the EU" (TFEU). |
| VDGH | Verband der Diagnostica-Industrie, e.V. (German Diagnostics Industry Association); national economic body which represents the interests of the diagnostic and Life Science manufacturers in Germany. |

| ZLG | Zentralstelle der Länder für Gesundheitsschutz bei Arzneimitteln und Medizinprodukten (Central Authority of the Länder for Health Protection with regard to Medicinal Products and Medical Devices) |
| --- | --- |

**Glossary**